DATE DUE

NO 12 '96			
FE 18 '96			
JE 2 9			
DE 11 9			
MY 22 00			
AG 3 00			
MR 30 01			

DEMCO 38-296

AUTOMOBILE DESIGN
Twelve Great Designers and Their Work

Second Edition

SAE Historical Series

Edited by

Ronald Barker

and

Anthony Harding

Published by:

Society of Automotive Engineers, Inc.

400 Commonwealth Drive

Warrendale, PA 15096-0001

at *Designers and Their Work* by David & Charles
eat Britain.

Library of Congress Cataloging-in-Publication Data

Automobile design : twelve great designers and their work / edited
by Ronald Barker and Anthony Harding.—2nd ed.
    p.   cm.
    Rev. ed. of: Automobile design / Ronald Barker. 1970.
    Includes index.
    ISBN 1-56091-210-3
    1. Automobile engineers—Biography.   2. Automobiles—
Design and construction—History. I. Barker, Ronald.
II. Harding, Anthony. III. Barker, Ronald. Automobile design.
TL139.A98 1992
629.23'1'0922—dc20
[B]
                                   91-43599
                                       CIP

Copyright © 1992 Society of Automotive Engineers, Inc.

ISBN 1-56091-210-3

CONTENTS

LIST OF
ILLUSTRATIONS

LINE DRAWINGS IN THE TEXT

INTRODUCTION

THE IDEA BEHIND this volume germinated in the late '60s as a proposal for a series of magazine articles, each to be written by an expert on the particular subject; but how could one compress the character and achievements of a man like Ferdinand Porsche within such narrow confines? So I then approached Anthony Harding, who had been responsible for the publication of many good books on motoring over the previous twenty years. Together, we drew up a list of prospective subjects, deliberately omitting some whose work had already been widely recorded while including others less well known but perhaps more colourful to deepen the perspective. It would have been presumptuous and contentious, in any case, to have picked out those whom we personally considered the greatest or most significant. Men like Adam Opel and Wilhelm Maybach of Germany, Henry Ford and the Stanley brothers of the USA, Herbert Austin, W.O. Bentley and Georges Roesch of England, Louis Renault and Ettore Bugatti of France, Vincenzo Lancia of Italy and many more, would certainly have qualified under this title.

Our objective was to portray designers as people, where possible giving as much emphasis to their backgrounds and characters as to their careers and creations. But in some cases any amount of research could produce little about the man himself—maybe because he was colourless or reserved as a personality and expressed himself lucidly only through his work, or perhaps because the years had swallowed up his associates and friends and no one could be found to provide personal reminiscence or open up family records.

This is a truly international collection, including a British subject of Greco-Bavarian origins born in Turkey, a Swiss whose cars were manufactured in Spain, France and under license in Czechoslovakia, an Austrian whose working life was spent mostly in Czechoslovakia, and a Bohemian who became a Czech but whose product proliferated in Austria and Germany, and for this new edition we have added an Italian, Dante Giacosa, whose creations have also been produced in factories far beyond the bounds of his own country, including segments of the now disintegrated Eastern bloc.

I noted when the First Edition was prepared that, of the two American authors (Borgeson and Sloniger), one has a Lebanese wife and lived in France, the other a Czech-born wife and a home in Germany. Our Dutch author, Kousbroek, with his American wife, is domiciled in Holland, and the late Jacques Ickx represents Belgium. Hendry is a New Zealander, de la Viña Spanish, and the rest of us are English. With fourteen writers (including two collaborating pairs), there was no question of maintaining any consistency in style or material even had this seemed desirable, so no unitary thread joins one chapter with the next.

When you have read the chapter by Jacques Ickx, that master of painstaking research and analysis complemented by philosophic deduction, you may come to realise that father Bollée was the true father also of the motorcar as a means of independent transport for individuals. He was by no means a pioneer in the use of steam for road transport, but he did foresee the need for relatively small, rapid vehicles in which people could drive themselves and their families about the country, where they wished and when they pleased, without dependence on set routes and timetables. He was as much concerned with controls and controllability as with the power plant, applying his thoughts to new suspension techniques and steering systems, to versatile and trouble-free transmissions, and to bringing the entire operation of the vehicle, other than stoking the fire, under the control of the driver. Every one of the Bollée steamers had independent front suspension of one sort or another. He tried front engines, back engines and twin amidships engines. He even built a steam road-and-rail track car with four-wheel drive, four-wheel steering, all-independent suspension—and even a steam servo to assist the clutch and steering. As with so many other prolific and far-sighted inventors, he was years ahead of his time, and I share Jacques Ickx's view that his work has never received the appreciation and recognition it deserves.

Who else qualified to write about Frederick Lanchester than the late Anthony Bird? Although he never met the great 'Dr Fred,' for many years he enjoyed a close friendship with brother George, who helped to build the first car to bear their name in 1895, and for some years he ran one of their 1903 tiller-steered cars—the one with two crankshafts and six connecting-rods for its two pistons. Dr Fred paid little heed to what other manufacturers were doing. He worked things out for himself from first principles and, like Bollée, was just as concerned with the frame structure, balance and manageability of his machines as with their means of propulsion. His designs were initially too unconventional for wide public acceptance, and subsequently became a bit outdated and 'old hat.' Although he backed out of direct car manufacture around 1910 and applied his intellectual genius to other sciences, he continued to contribute to automobile engineering through papers put before the Institution of Automobile Engineers (which eventually amalgamated with the Institution of Mechanical Engineers to become the Automobile Division of that august body), as well as other professional and academic societies. The current state of the art owes a lot to the knowledge which other designers have absorbed from him.

Maurice Hendry, a New Zealander, made a particular study of American automotive history. His subject, Henry Leland, is not widely appreciated outside the USA, but his influence on the American industry as a whole was very great. When he helped to found the Cadillac company in 1902 he was already sixty years old, and had a standing reputation as a designer and builder of precision tools. He had learnt about close working tolerances and interchangeability of parts in the small arms business, and many of the pioneer motor builders came to him for help and advice. He might be said to have done for the rich what Henry Ford did for the poor, churning out luxury cars by the thousand instead of the hundred, and his V-8 Cadillac of 1914 was certainly one of the great American classics. Although after his time there was a vogue for the straight-eight, through the 1920s and '30s, his Cadillac can be considered the true and direct ancestor of the American V-8.

Hans Ledwinka, who died in 1967, is perhaps little more than a name to most automobile historians outside his own sphere of operations in Austria and Czechoslovakia, but it would take a book to show the full range of his activities and thinking. It was very tragic that he died in a motor accident not long after being released from a Czech gaol after six years' imprisonment on trumped-up charges, and before Jerrold Sloniger was commissioned to write his story, so that they never met. However, Ledwinka's son, Erich, provided some personal background, and after a formidable research job Jerrold has put together a perceptive analysis of his work. Ledwinka applied his inventive genius to engines, transmissions, chassis structures and suspension systems with equal fervour and ingenuity. His thinking had much in common with Ferdinand Porsche's, and had he operated closer to the heart of European motor manufacture he would surely have expanded his influence and gained wider recognition.

Despite the almost legendary status of the Hispano-Suiza car, it proved almost impossible to bring to light much about the nature and personality of Marc Birkigt, the individual behind the whole enterprise from its foundation in Spain in 1904 right through to the last of the cars in 1938. It seems that, away from his work, he was a family man who moved socially in the top bracket, but beyond his own close professional circle he was content that the world at large should know him only through his products, seeking no personal lip-homage or publicity. Harding and I ran into several editorial *cul-de-sacs* before pinning down Michael Sedgwick and José de la Viña to collaborate on a combined operation, each having special knowledge and research facilities to bridge the geographical distance between Britain and Spain.

Marc Birkigt is remembered particularly for the *Coupe de l'Auto* voiturette racers of 1909-12 and the Type Alfonso production sports cars (among the very first) developed from them, all with long-stroke T-head four-cylinder engines and built in Barcelona; and for the superb luxury cars he built in France from 1919, while the Spanish factory turned out less sophisticated models. He was primarily an engine man and, like Royce, he did his best work in the aviation world although, unlike Royce, he also provided his motoring customers with overhead camshaft engines for some years. He was no innovator in chassis design; true, the 32CV of 1919 had the world's best brakes (on all four wheels, with mechanical servo) of that period and probably the lightest steering of any large car, but Birkigt never indulged in adventures away from channel frames and beam axles on leaf springs, and his 1938 cars were structurally much the same as the 1919 ones underneath their new-style bodies. Anyone with personal experience of the big V-12 of 1931-8 would probably agree with me that this superlative engine was served by a very indifferent transmission in a chassis that also did it less than justice.

In contrast, Dr Ferdinand Porsche disseminated his creative talents over the broadest imaginable technical fields—touring, sports and pure racing cars, petrol and diesel aero engines, electric and petrol-electric cars and haulage vehicles, military tanks, diesel trucks and so on. Commercially, his *chef d'oeuvre* has proved to be the Volkswagen 'Beetle,' the most prolific seller since the Model T Ford (of which over two million were made in one year). In fact, the Beetle surpassed the Model T's record of something over fifteen million total production. Like Ledwinka, Porsche suffered imprisonment (by the Allies) following the Hitler war, when a septuagenarian, but was released after nearly two years. The Herculean task of exploring the many facets of Dr Porsche's work, putting them into perspective and compressing them into a single chapter was entrusted to D.B. ('Bunny') Tubbs, who not only knows his motoring history but also has a way with words that helps to make it digestible.

Three of our designers are associated primarily with competition cars—Harry Miller, Vittorio Jano and Colin Chapman. Although the products of Miller and Jano were contemporary for many years, there was the breadth of the Atlantic between them in style and purpose, and whereas these two men are remembered chiefly for their engines, Chapman was a master of the chassis structure and suspension geometry, who bought his engines from proprietary sources.

Griffith Borgeson rates Harry Miller 'the greatest single figure in American motor-racing history, as well as one of the greatest automotive designers that country ever will produce.' The Miller story is one of extreme ups-and-downs, shattering success followed by dismal commercial failure and success again, but culminating in long, sad years of poverty and depression and painful illness. You can detect in Borgeson's writing that Miller is for him an idol, to be worshipped for the things he made and the way he made them, and to be revered as a character with all his faults and idiosyncrasies freely accepted. It is an intensely moving story, put together here with the same meticulous intensity that its subject applied to his racing engines, his centrifugal blowers, his front-drive and four-wheel drive transmissions. For a very long time the four-cylinder Miller racing marine engine, carried forward and developed by Harry Miller's one-time associates, Fred Offenhauser and Leo Goossen, remained a favourite powerplant for the Indianapolis 500.

You really have at least to begin understanding Gabriel Voisin as a man before you can understand his cars, and his two books, *Mes 10,000 Cerfs Volants* ('My 10,000 Flying Stags') and *Mes 1001 Voitures* are among the most entertaining and intensely personal autobiographies ever published. Probably the historians will remember him primarily as a pioneer aviator and aeroplane-maker. The cars that came later (from 1919) were native, domestic French as distinct from conformist European, and sometimes idiosyncratic to the point of eccentricity. From his own writings and utterances it is clear that Voisin rather enjoyed being thought controversial and cantankerous, forever pricking other people's balloons, but being extremely articulate he has also expounded on the philosophical reasoning and logic motivating his own work. Is he truly a 'Great Automobile Designer'? Probably yes—but if not, certainly 'Great, and an Automobile Designer,' in the sense that he is among the more prodigious characters thrown up by the industry. Rudy Kousbroek, who wrote the Voisin chapter, knew and understood the man, being himself a bit of an academic who has written books and essays about logic and the history of science and techniques and such things. Voisin died only a few years after his chapter was written, but his creative life was already over and so it is retained word-for-word in this new edition, as it was written.

My first encounter with Sir Alec Issigonis was a few weeks before the Mini's launch on 26 August 1959, while preparing for a round-the-Mediterranean PR trip that an *Autocar* colleague and I had persuaded our editor would be good for his circulation! On that very date we were duly waved away from London's Festival Hall by racing champion Jack Brabham—the only occasion I ever saw Jack confined to a wheelchair. Thereafter I grew to know Alec quite well, and so was able to gather material for my piece about him in the tranquility of his home, armed with a tape recorder. In retrospect, the limitations of this approach may be outweighed by the seeming advantages. Anyhow, his conversation was so stimulating, so rich with provocative aphorism and tongue-in-cheek comment that I decided to include brief verbatim extracts from these tapes.

Like any human being, Sir Alec was open to criticism and was openly criticised. When things appeared to go badly, the critics would pounce as vigorously as the sycophants would adulate when he was balanced on the crest of a wave of success. The same happens in all professions, but in the automobile world engineers are mostly concealed behind a screen of industrial reserve that discourages individuals from becoming public figures and taking the brickbats as well as any indulgent recognition that's going.

In appearance, Issigonis put one in mind of an amiable bird of prey, with that slight stoop and the aquiline nose between penetrating blue eyes, and as he talked the large, expressive hands (of an inspired blacksmith, as he put it) were never still. He described himself, with an understandable and frank touch of conceit, as 'the last of the Bugattis.' This is not to suggest that there are not, nor will be, other creative designers with a touch of genius; only that they are less likely to be given so much personal rein inside a big organisation, or named and 'released' for public appraisal. But I wonder: When another Issigonis crops up, could his extrovert nature really be suppressed?

Then we have Colin Chapman. His biographer, Philip Turner, was for many years Sports Editor of the British magazine *Motor,* so had ample opportunity to get to know the man and his Lotus cars. He has now taken the not altogether happy task of writing the final pages of Chapman's life.

Chapman was at his zenith when the original chapter was written, and Turner then described him as the greatest living designer of racing cars and of high-speed road cars. At the time I questioned whether one could apply such absolute terms to him without qualification. He had never designed an engine—and the engine is a rather important part of a racing car. And while his road cars had the exceptional dynamic qualities one came to expect of a Chapman design, their supremacy was certainly not on all counts or over all other categories of the type.

Chapman was primarily a structures and suspension king, stemming from early training in the aircraft industry; with his knowledge of stressing he was able to make his cars lighter than most for a given strength (although, heaven knows, there have been some breakages!). Time and again he led the way for others to follow—witness, for instance, the first monocoque GP car (Type 25) of 1962. Had he not become so successful as a manufacturer and instead persisted as a competition driver, most likely Chapman would have reached the very top bracket in that profession.

Two decades have passed since the first edition of this book was published (in England, by David and Charles). In that time three of its living subjects, Gabriel Voisin, Sir Alec Issigonis, and Colin Chapman, have died. Whereas Voisin had long retired and Sir Alec was on the threshold of retirement in 1970, Chapman was in his prime, and who could have guessed that his particular strand of genius would be so abruptly severed? Fortunately Lotus cars have been able to survive without him, but not without passing into Transatlantic hands.

After 1970, the Issigonis saga developed into something of a personal tragedy, with professional rejections and degenerating health that progressively restricted his social activities and circulation. It's the more ironic, therefore, that the Mini, brainchild of this once vivacious and ubiquitous figure who lived into virtual obscurity and pathetic old age, should remain in such international favour more than three decades since its inception.

The first edition should have covered a round dozen subjects, and that one of them should have been Fiat's Dr Dante Giacosa. For some reason, most probably concerned with time and money, we had to make do with eleven. Well, here is now, one of the "greats" for whom I have always entertained a deep respect and whom I was privileged to meet several times when he was at the top of his profession and I had been sent as a technical writer to survey his latest creation. I recall him as accessible, charming, debonaire, without conceit. Leonard Setright's erudite evaluation seems a fitting *bonne bouche* to complete a 12-course indulgence.

In the hundred-odd years between Bollée *père's* first steam brake *L'Obéissante* and Chapman's Lotus 72 GP racer, a lot of ground has been covered in both senses; yet one can still link them with a few common features. For instance, four wheels, independent front suspension, amidships 90 degree V engine(s)—Chapman's with eight cylinders in one unit, Bollée's with four cylinders in two. Over now to Jacques Ickx at the Bollée bell foundry in Le Mans . . .

Shorncote, Cirencester, 1991   Ronald Barker

1 THE BOLLÉES

by JACQUES ICKX

AMÉDÉE BOLLÉE PÈRE (1844–1916). French: Designer-constructor of a succession of individual private steam vehicles including *l'Obéissante* (1873), *La Mancelle* (1878), *La Nouvelle* (1880) and *La Rapide* (1881). His technical innovations included independent front suspension, a differential gear for the back axle, and front, rear and amidships engines. His elder son, AMÉDÉE FILS (1867–1926), made petrol cars from 1896–1914, including a streamlined racing machine in 1898. LÉON (1870–1913) is especially remembered for his three-wheeled tandem of 1895, but manufactured conventional motorcars from 1903.

AMÉDÉE BOLLÉE PÈRE

IT WOULD BE inconceivable to leave the story of the Bollées (the father and his two sons) out of a book about great automobile designers. All three were fertile inventors, untiring pioneers and creative builders of an industry, but what really sets them apart most from other pioneers is the amazing quality of their design work. More than a century afterwards, the balance inherent in their creations and the neatness of their forward-looking solutions still make a striking impression on present-day observers.

Most people know that the Bollées brought fame to the town of Le Mans, where their direct descendants (they, too, are inventors) still run the piston-ring factory established after the First World War by Amédée Bollée *fils*. But their origins and past were elsewhere, in that part of the countryside of old France which borders the banks of the Haute-Marne and is known as le Bassigny. This was the country of the *saintiers*, or itinerant bell-founders. Each spring they would start off on their annual trek, setting up their workshops wherever there was a bell to be cast (for in those days there was still no means of transporting such heavy objects).

The French Revolution was to change many things, and the advance of technology brought even more changes. The *saintiers* were to disappear. One bell-founder called Ernest-Sylvain Bollée, who had taken up the craft in 1830 at the age of sixteen, was among the first to realise the possibilities opened up by the railways. In 1839 he set up a permanent workshop at La Flèche, near Le Mans, and three years later settled in the town of Le Mans itself, in the Avenue de Paris. The foundry was soon to win a worldwide reputation.

Ernest-Sylvain married and had three sons. In his wisdom, he wanted them all to work on their own while he was alive and to inherit a business of their own when he died, and it was for this reason that he went on to set up a factory to produce hydraulic rams and another to make windmill pumps. Following ancestral tradition, the bell foundry was the concern of the eldest son, Amédée, who was born in 1844 and was to be known to the motoring world as Amédée Bollée *père*.

We have a clear picture of the personality of Amédée Bollée *père* from numerous writings and various items of evidence which have been piously preserved. To say that he was a fine man would not be enough— by today's standards he seems to have been goodness and honesty personified; good and honest to the point of candour, if not indeed of naïveté. A profound idealist, he was devoted to his father, and it was this devotion that produced the strange quirk which was to turn him for a while into an automobile designer. He needed to equal his father, but without surpassing him; and this is why, while carrying on his father's work in the foundry, he wanted in his turn to create a fourth business. It must be said at once that it proved to be a cruel and bitter experience from which he emerged so bruised and battered that he abandoned his own work at the very moment it might have succeeded. Instead, he returned full-time to the occupation of bell casting (which he had never given up), and to an almost ritual perpetuation of his father, whom he no longer sought to equal, far less to rival.

Bell casting is an advanced scientific technique, a marvellous combination of metallurgy with the worlds of shape and sound. It has provoked innumerable treatises and historical studies, widely read by fanatical amateurs as passionate in their devotion as any car-lovers. In this world of bells, Amédée Bollée *père* is an historical figure; he did a great deal of creative work and occupies such a leading position in this particular branch of technology that specialists in that line seldom imagine he could have done anything else. Indeed, Amédée Bollée *père* himself would undoubtedly have been vastly surprised had he been told that future generations would honour him as anything other than a bellfounder.

There is another aspect of Amédée Bollée *père* that should be noted: he went to school up to the age of ten or twelve at most, and then straight into the foundry to learn direct from his father what he needed to know to carry on the craft. That the training was effective was shown in 1859 when Ernest-Sylvain fell gravely ill and his fifteen-year-old son successfully took over the running of the foundry. A few years later, applying the scientific aptitude of a fresh mind to the practical experience of centuries handed on to him by his father, the young man who was to be Amédée Bollée *père* laid the foundations for new knowledge by working out a rational method of analysing the components making up the timbre of a bell. Exactly one year later he was to apply this same technical bent to what was for him an entirely new field, and one in which he was even to believe that he was taking the first steps.

It is difficult in this day and age to picture how, barely a hundred years ago, men lived almost totally enclosed within their own times. There was then no dissemination of technical or scientific literature, and it was the great exhibitions of those days which were the chief means of spreading information. Thus it was that the Paris World Exhibition of 1867 brought two revelations to Amédée Bollée who, residing at Le Mans, 140 miles from Paris, had had little previous idea of what was afoot in the capital, and far less of the earlier achievements of the British horselesscarriage builders.

Page 25 Father and Son: Amédée Bollée *père* (1844–1917), the inspired designer of successful steam cars, with Amédée Bollée *fils* (1867–1926), an early student of aero-dynamics, and best remembered as an artist ever dissatisfied with his own work

Page 26 (above) Bollée's *La Mancelle* of 1878 set the classic pattern of the automobile with front-mounted engine, transmission beneath the passenger compartment, rear wheels driven through a differential, and independent front suspension; *(below) L'Obéissante*, 1873, was a compact and well-proportioned vehicle. *Père* Bollée probably pioneered the flush-sided body on a true perimeter chassis frame, as well as independent front suspension

Page 27 (*above*) Léon Bollée in the driving seat of his three-wheeler, *La Voiturette*, in 1896, with his younger brother Camille in front. An English industrialist paid 20,000 gold livres for the rights to manufacture this vehicle in the United Kingdom; (*below*) this pioneer example of streamlined bodywork, the *Soucieux* 'torpedo-boat', was built by Amédée Bollée *fils* in 1899 and the same design was applied to the aluminium-bodied cars which later took part in the Paris–Amsterdam run

Page 28 (*above*) Amédée Bollée *fils* at the wheel of his Model E, in 1907, with his father beside him. The price of this car was then almost identical to that of the contemporary Rolls-Royce; (*below*) a view of the factory which Leon Bollée established in 1900 at Sablons, near Le Mans. In the foreground is a chassis of the famous six-cylinder Type 30/45

The two revelations which came to Amédée Bollée were the Michaux velocipede, which he adopted at once, and a steam omnibus by Guidez and Colin, which was exhibited alongside tractors and road trains, and of which we know today only that its engine was at the rear. These two phenomena fused in his mind and, as he was to relate subsequently, he at once had the idea of building a private steam carriage—which he first saw as a means of giving the advantages of cycling to people who were too old for the sport. However, his inspired plans went further than this, and opened a door into the future.

Unlike the motoring pioneers who were to come on the scene twenty years later with the prime objective of making their mechanical carriages move at much the same speed as a horse-drawn carriage, Amédée Bollée in 1868 and at the age of twenty-three, made up his mind to build (in his own words) 'a fast private carriage'; and from that moment the course of all his future work was set.

From the very fact that he intended his carriage to travel at speed, Amédée Bollée could not be satisfied with the tricycle layout, nor with the front axle swivelling around a single pivot, nor with the driving wheels fixed solidly to the same axle (as they were on the road locomotives). He realised at once that he would need to provide steering which allowed the wheels to point at different angles, and that it would have to be possible for the driving wheels to travel different distances within the same time. In fact, solutions to these two problems were already slumbering in the files of the Patent Office, forgotten by everyone and unknown to Amédée Bollée. Lankensperger, a Munich coachbuilder, had designed such a steering layout in 1816 (for horse-drawn carriages), and in 1828 Pecqueur had planned the adaptation of the clockmaker's differential movement to locomotion. Bollée, however, found different solutions. Instead of using Lankensperger's simple quadrilateral system of rods, he steered his wheels (mounted in pivoted forks like those of a bicycle) by means of a chain running round a set of elliptical cams. And instead of a differential, he planned to drive each wheel by an independent engine with a single steam supply to the two engines, the pressure being automatically distributed in proportions matching the resistance offered. (See drawing, p 31.)

The elliptical-cam system of steering is, in theory, ideal and no more precise steering has ever been designed; but the lateral weakness of the forks and the inevitable stretching of the chain (despite the use of tensioners) rendered any further development impossible. Bollée added to the automatic distribution of tractive power an extra control pedal which completely cut off the supply to one or other of the engines, so making the carriage pivot round on one of its rear wheels, the corresponding front wheel then being at right-angles and the radius of the turning circle being no more than the diagonal of the carriage's own wheelbase and track!

A third realisation by Bollée was that although the driver of a fast carriage must inevitably be seated at the front, this was no reason for depriving him of control over the engine (as was the case with the steam carriages of the preceding era), even though it might mean introducing the complications of remote control. Thus it was that the 'cockpit' of the modern motorcar came into being. Having decided this, there was a fourth principle which guided Bollée's creative efforts—the engine which was taking the place of the horses, and which must necessarily be carried on board the vehicle, should not interfere with the passenger accommodation. Finally, Bollée evolved a fifth principle: the vehicle he had in mind must be neither a carriage fitted with an engine, nor a road locomotive fitted with seats—it had to be a completely new concept, matching his own requirements.

A variety of circumstances, including the Franco-Prussian War of 1870, set Amédée Bollée's enterprise back by five years, and it seems reasonable to assume that the maturity of execution evident in the eventual design, and which strikes the observer at first glance, was due to this delay. Here is none of the experimental and incongruous assembly of elements which so many others had and were still to produce. L'Obéissante (The Obedient One), which began to take shape in 1872 and was finished early in 1873, was a vehicle which was 'all-of-a-piece', compact, and a perfectly balanced and completely functional entity. It is surely true to say that if the most brilliant of modern engineers were restricted to the resources available at that time, he would have been unable to improve on the concept or the design. The vehicle is clean, logical and marked by a kind of animal elegance. Eighty-five years before the Mini, L'Obéissante was already using four-fifths of its overall length for carrying its passengers, and offering its driver the advantages (which, though probably still unsuspected, would soon become evident) of independent front suspension a full half-century before the Sizaire brothers.

Bollée achieved this masterpiece of passenger accommodation by using twin-cylinder engines in a 90 degree vee conformation, housed comfortably in the space between the two wheels on one side, underneath the fore-and-aft bench seat for the passengers. This kept the vehicle short, with its floor at a reasonable height above the ground. The two independent, two-ratio transmissions had the effect of offering a third gear (by having one running at high speed and the other at low speed). We may note the choice of double-acting cylinders in a 90 degree layout;

with this arrangement the balancing of inertia forces is better than that of a modern in-line four-cylinder engine, and the evenness of the torque equals that of an eight-cylinder four-stroke.

The vehicle was remarkable for having what we should now term a full perimeter frame (of riveted iron) arched over the iron-spoked wheels and flush with the metal panels of the covered brake wagonette coach-work. Each front wheel was straddled by two long and supple, full-elliptic springs in an independent fork. The elliptical cam steering gear, too, was fully suspended. The coachwork, in the style of a covered wagonette, was again in metal and fully panelled. While the central

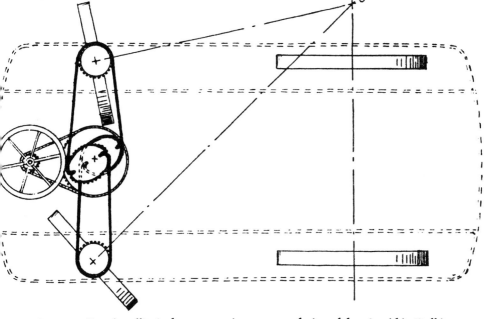

Theoretically, the elliptical cam steering system designed by Amédée Bollée père for his L'Obéissante gave geometrically precise steering up to very extreme angles; in practice, however, the chains stretched irretrievably

seat provided for the driver was ideally arranged for his comfort, it did unfortunately complicate access to the vehicle for the passengers through a single entry at the front end, and this was the weak point of the design. On the other hand the 'cockpit' was a model of its kind, with a large horizontal steering-wheel, a brake-wheel (also horizontal) on the right hand, levers for speed control (accelerator) and reverse or cut-off beneath the hand, the two gear levers on the floor to right and left, and two pedals for opening or closing the supply to one of the engines. L'Obéissante would carry up to twelve passengers in addition to the driver and fireman, weighed over 4½ tons in road trim, had a range of 15½ miles (after which one of the engines was used to actuate a suction pump which replenished the water tank from a roadside pond or stream), reached a speed of 25mph and could climb a 1-in-20 incline. With *L'Obéissante*, the steam carriage not only ceased to be a far-off dream but straightway met the requirements of practical use.

The next step was to ensure reliable operation of the machine, and to carry out a patient process of improvement until it satisfied its creator. Amédée Bollée was painstaking and hard-to-please, and though the wheels of *L'Obéissante* first turned in January 1873, it was October 1875 —after two-and-a-half years of almost daily sorties—that the first steam-carriage designer considered his work was complete; he decided to show *L'Obéissante* in Paris, where reputations were made and successes born. His acclaim was enormous: after a faultless journey by road, he drove along the boulevards day after day, taking important people for rides, astounding the crowds and attracting attention from the press which found an immediate response among the public.

It may have been because this was described, for the first time, as a vehicle for pleasure trips; it may have been because of the level of performance reached. At all events, Amédée Bollée found an impressive mail awaiting him when he returned to Le Mans a fortnight later. For the first time the urge to own what was one day to be the motorcar was spreading and making itself felt, and everyone wanted to know more about *L'Obéissante*. It seemed that success was to come at the first attempt; but Bollée's faith was to suffer some shattering blows. All these enquiries, and all this exchange of letters, was not to bring in a single order for the private mechanical carriage he had produced and whose possibilities he had demonstrated, and his amazing feat of creation resulted, in the end, in an order for two tramcars.

Was Amédée Bollée before his time? The response of the public seems to suggest that this was not the case, and it is more likely that his work was presented to the world too soon. The public were enraptured by the doings of Bollée the artist, but was well aware that this was the work of an artist.

Forced back into the field of public transport, on which he had sought to turn his back, Amédée Bollée produced some remarkable new designs. One of the tramcars was ordered by his friend Dalifol, who wanted to propose it to the Compagnie des Omnibus de Paris, and this in itself at once presented an extremely complex problem. As the horse trams of that time were small and had a swivelling front axle, tram-tracks could be given tight curves of a seemingly impossible radius. Moreover, the track was a single one for use in either direction; when two trams met, the horses of one of them tugged sideways on the central shaft and pulled the tram out of its rails, continuing along the roadway until the other vehicle was clear. This is what the driver of the Bollée-Dalifol tram would have to do with a double-decker vehicle carrying fifty passengers (since only a high carrying capacity would make mechanical power worthwhile).

Bollée saw at once that he would need to have both *steering and traction on all four wheels*, and since he wanted independent suspension on all four wheels as well, the problem was a real poser. This time he produced (forty-five years before its time) what was to be a distinctive suspension system for the future Lancia Lambda—a telescopic 'sliding-pillar' with a coil spring. For the steering he re-invented, without realising the fact and in a duplicated form acting on both sets of wheels, the Lankensperger layout; however, he added a characteristic extra feature in that the steering of the front and rear pairs of wheels could be made independent of each other. This arrangement allowed the tram to leave the rails with all four wheels at the same time, move sideways during this manoeuvre (with a minimum of disturbance to other traffic), and then regain the rails, again in a single movement.

The problem that remained, having solved those of suspension and steering, was to drive the wheels. Here, Bollée repeated the basic layout of *L'Obéissante*—two independent engines in a vee conformation, housed lengthwise under the benches between the two pairs of wheels. This time an extension on the end of the crankshaft drove a chain transmission located along the centre of the tramcar, this in turn driving the front and rear wheels on the same side of the vehicle through other transverse shafts. The latter had a double universal joint and sliding links so as to follow all the wheel movements.

It will be seen, therefore, that it was Amédée Bollée and not the Marquis de Dion who first invented a transmission using transverse, universally-jointed shafts. It was he, and not Latil, who first invented front-wheel drive. It was he, and not Sizaire, who first designed a vehicle with independent suspension on all four wheels. Nor does this complete the list—for the Bollée-Dalifol tramcar had a power-assisted clutch and steering as well! For each of the latter, all the driver had to do was to open a cock which applied steam to one or other end of a piston whose movement engaged the drive to, or steered, the wheels, as the case might be.

The Bollée-Dalifol tram underwent official trials on 15 November 1876, travelling from La Villette to Belleville at an average speed of 15mph, and reaching a top speed of 27½mph. Nevertheless, it was considered too heavy, and was rejected. And Bollée could go back to his real love, the private steam carriage.

The commercial failure of *L'Obéissante* made Bollée think about his design from the psychological rather than the technical angle. Unfortunately for him, he was not willing to see the whole truth; which is understandable, since if he had been willing to realise that steam power (the only form of power possible at that stage) was not suitable for driving a private carriage, the only answer would have been to give up the idea altogether. So Bollée did not question the principle of the steam engine, but that of the design of *L'Obéissante* itself: why had it not proved acceptable? He found three shortcomings, which appeared significant to him long before the others.

First of all, *L'Obéissante* was ungainly—to prove attractive, a steam carriage needed to look as much like a horse carriage as possible. Next, there were the fore-and-aft benches, on which the passengers automatically turned to sit crosswise as soon as the vehicle began to gather speed; the next design would need to have transverse seats. And finally there was the ridiculous single entrance up a narrow ladder, which forced the passengers to squeeze past the driving seat impeding their access. The steam carriage could only be comfortable if all the seats were directly accessible by mounting a simple step. To bring these three features together, Bollée turned to the outline of the low, four-wheeled 'victoria', although this meant entirely re-thinking the mechanical design.

Clearly the boiler must be placed at the rear, for the comfort of the passengers. The engine could then no longer remain in the middle of the vehicle, but would have to be moved forward. As there was insufficient space beneath the driving seat, it would have to be further forward still, even to the extent of overhanging the front wheels; and the engine would then need a cover of its own (something still quite unheard of), which would come to be called the 'bonnet'. However, there could be no question of using chains to couple an engine at this distance from the rear wheels—and these would have to be the driven wheels, from the mere fact that they were carrying the weight of the boiler which would lend them adhesion. Following on the forward position of the engine, the shape of the victoria thus led to a transverse mounting which would allow the transmission to run direct from the end of the crankshaft to the rear of the vehicle. And to transmit differential movement to two wheels being driven by a single shaft, the only answer was, like Pecqueur (whom Bollée did not know), to adapt the watchmaker's differential for locomotive purposes, giving it its final form as used later for the motor-car (ie, combined with a bevel gear). In short, all that was needed for *La*

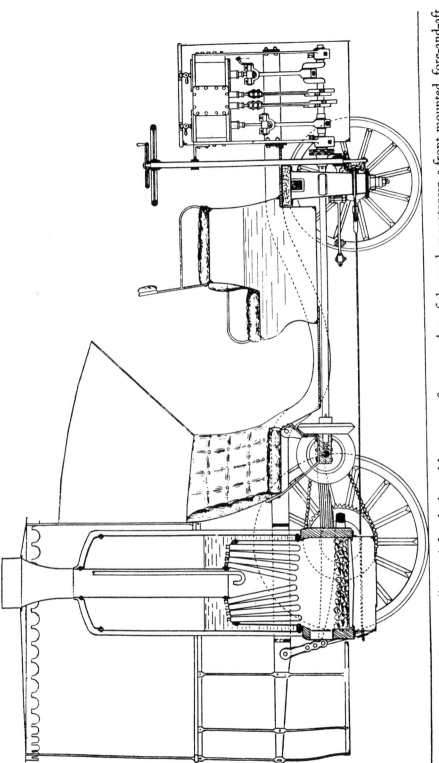

The general layout of *La Mancelle* (1878) foreshadowed by twenty-five years that of the modern motorcar, a front-mounted, fore-and-aft engine and a transmission shaft in line with the crankshaft driving a differential, via a tapered pinion, to supply power to the rear wheels

Mancelle (Miss Le Mans) of 1878 to have the general features of the classic motorcar was a gear-change system; and this, too, was to be incorporated at a later date and—following the same inevitable logic—was to form a single unit with the engine (thus constituting, in 1880, the direct drive later attributed to Louis Renault!).

All this would have been remarkable enough: but *La Mancelle* introduced something else—a front suspension system so progressive that it has remained a feature of automobile design almost until the present day. With this vehicle Bollée achieved independent springing of the wheels by the simplest and lightest means possible: two transverse laminated springs one above the other, joined at the ends by the wheel pivots and constituting a flexible axle. To provide fore-and-aft stiffness, Bollée made the bottom spring a single leaf of substantial hexagonal section. However, since the steering pivots would now be moving up and down, the simple quadrilateral system of steering rods used on the Bollée-Dalifol tram was no longer adequate, and Bollée substituted for it the first-ever example of a split tie-rod, with the two halves ending in articulated joints so as to allow for the vertical movements of the wheel. (See drawing, p 38.) The system as a whole was unequalled at its time, and all the thirty-two Bollée vehicles later built at Le Mans or (under licence) in Berlin were fitted with it. It then sank into oblivion, and Sizaire was to make his name by rediscovering the layout twenty-seven years later.

We should not overlook, either, the general advance in Bollée's engineering during the five years which separate *L'Obéissante* from *La Mancelle*. The latter weighed under 2¾ tons; it needed 12hp instead of 20 to reach a speed of 22mph and, in particular, to achieve average speeds close to its maximum when travelling through hilly countryside. With *La Mancelle*, stopping and restarting on a 1-in-22 gradient presented no problem.

But perhaps the most remarkable aspect of all is that Bollée was able to offer *La Mancelle*, prepared for the Paris World Exhibition of 1878, at a price of 12,000 francs, while the antiquated Randolph carriage exhibited alongside it cost 45,000 francs to build. Was the pioneer of the private carriage—ie, of the future motorcar—to achieve his goal? No: for he found only one customer for his steam victoria, and what he was asked for once again was passenger transport vehicles derived from his design. It was these vehicles, too, which the Wöhlert factory was to build when (under doubtful circumstances which it would take too long to go into here) it obtained the use of Bollée's patents.

For Amédée Bollée, however, the giant (100hp) tractor which he supplied to the Société Metallurgique de l'Ariège was to lead to fresh discoveries. He was, for example, to make the crankshaft of the in-line twin-cylinder engine of this giant version of *La Mancelle* act as the main-shaft of a gear-train, with the layshaft situated beneath it; this, as long ago as 1879, was the layout now characteristic of the BMC Mini, adopted as a solution to the same problem of saving space. The transmission shaft, thenceforth horizontal, was aligned with the layshaft carrying three gears. Nor was this all; for this transmission shaft, after driving the differential, continued back to the rear end of the tractor where it was linked, through a large universal and sliding joint arrangement, to a further transmission shaft which in turn worked through a differential to drive the wheels of the tender (ie, the trailer wheels were themselves powered). Where this tractor (named the *Marie-Anne*) was concerned, this was as far as things went; but the patent provided for the trans-mission of power to all the wheels of a road train, thus anticipating by twenty-four years the famous Renard train.

Up to this time Amédée Bollée had been building his vehicles in the bell foundry, financed by a series of loans from his father (for his monthly salary as director of the foundry amounted to no more than 500 francs). At the end of 1879, stirred by the fantasies of a pathological liar who had appointed himself Bollée's general agent, he set up his own works (financed by a bank loan) 200 yards away from the foundry buildings and ran the two businesses side-by-side. No less than seventeen workmen were employed on building the steam carriages, and we should add at once that Bollée was never paid for the two giant tractors, nor for a forty-seater omnibus, nor for a replica of *La Mancelle*, which were the early output from his factory.

We can well imagine that while *La Mancelle* and the second tractor, named *Elisabeth*, were given further triumphal demonstrations in Austria and Germany (where his general agent was working up 'important orders', all of which fizzled out), the situation in Le Mans was becoming critical, and it was probably to help him that his father, Ernest-Sylvain, and his brother Ernest, became, one after the other, customers of Amédée Bollée. These orders enabled him to continue to develop his creative genius.

Ernest-Sylvain was a customer who knew just what he wanted. First of all, he wanted a closed carriage, like the Randolph shown at the Paris Exhibition. And he wanted the vehicle to incorporate his own recent in-vention (following the example of his son) of a single circular slide-valve for distributing the steam between two parallel cylinders; besides being simple, this system offered the advantage of opening the ports more quickly, thus speeding up the rate of utilisation, and of a greater expan-sion than was possible with the Stephenson slide-valve. On the other hand, he was quite happy with the general arrangement of *L'Obéissante*.

c

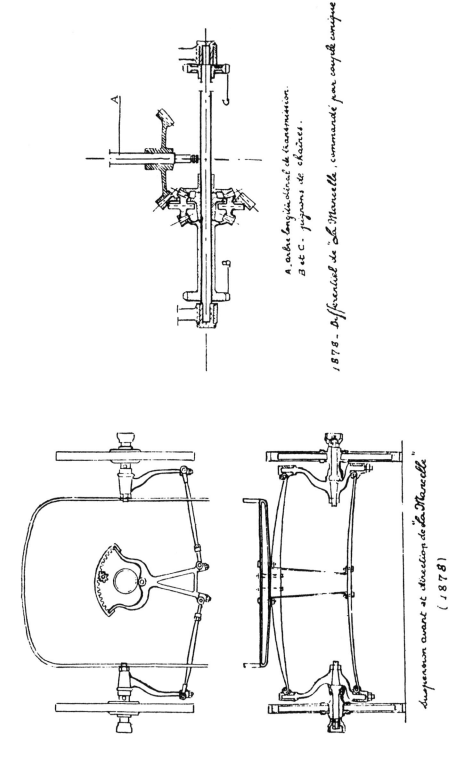

A. arbre longitudinal de transmission.
B et C. pignons de chaînes.

1878 - Différentiel de La Mancelle, commandé par couple conique

suspension avant et direction de "La Mancelle"

(1878)

The independent front suspension of *La Mancelle*, comprising two multi-leaf springs, was still in use on the last Dyna-Panhards in 1963. The divided trackrod steering, with joints at each end, was completely new, as was the tapered pinion drive to the differential which later became a standard design feature

Faced with this personal problem, Amédée Bollée produced what was to remain undoubtedly his design masterpiece.

From *La Mancelle* he took only the front wheel layout, which could not be improved on at that time. He used a very simple chassis, fitted his vertical Field boiler in the traditional way, immediately behind the rear axle, set his parallel-cylinder steam engine horizontally and mounted it under the chassis just in front of the driving wheels. This gave him a completely flat floor, on which he arranged at the front an omnibus bodywork with the three pairs of seats facing forward, an enclosed driving seat (on the right side), easy access through a door at the front, and curved windows giving this front end a rounded shape which would offer less wind resistance. The whole of the rear half of the vehicle was laid out to give the fireman easy access to the boiler and machinery. This design shows not only a truly functional approach and a superb general balance, but also for the first time a twofold concern with styling. The mechanical carriage takes on a shape and styling of its own, quite different from those of the horsedrawn carriage, and at the same time we can see a new interest in achieving a continuity of line. The sides of the platform for the mechanics follow the level of the waistline before dropping at the back, while the canopy above this platform is at the same level as the roof of the main bodywork.

La Nouvelle's transmission was, again, a masterpiece of ingenuity. The main gear shaft was attached to the end of the engine crankshaft (set transversally), and the tubular layshaft was combined with the structure of the differential, with the half-shafts to the wheels running inside it. The drive unit of *La Nouvelle* thus achieved an absolute record in compactness.

Turning to the general engineering aspects of the steam carriage, it is to be noted that *La Nouvelle*, which was longer than the *L'Obéissante*, weighed only 62cwt (7,055lb) in road trim against the latter's 95½cwt (10,580lb), used 6½lb of coal to the mile instead of 9, and 2½ gallons of water as against nearly 5; it had a range of 37 miles instead of 15½, could be brought up to working pressure in seventeen minutes instead of half an hour and could achieve much higher average speeds (up to 24mph) with a maximum of 28mph. This vehicle was thoroughly practical—provided, of course, that one understood how to cope with a steam engine. During the remaining eleven years of its owner's life, it was to cover almost 10,000 miles and its swansong was triumphant; put back into running order in 1895, a full fifteen years after it was built, it was to finish the course of the Paris–Bordeaux–Paris Marathon Race and be the principal rival to Emile Levassor.

When, in his turn, Ernest Bollée *fils* became his brother's customer, his requirements—since he was still a young man—was for something more than a mere means of transport. For him, mechanical locomotion must have a sporting image. The specification stretched Amédée Bollée's imagination to the full, and was to bring about a sensational move towards the vehicle—still inconceivable at that time—which we were subsequently to know as the motorcar. The fundamental novelty was the 'dismissal' of the fireman. The Field boiler (with automatic coal feed) had progressed to the point where the designer felt that supervision of it could be entrusted to a single driver-cum-mechanic. This meant that the boiler had to be shifted right to the front of the vehicle, giving it the unfortunate appearance of a small road locomotive. 'Unfortunate' is the right word here, for *La Rapide*, as it was to be called, was very far from being a locomotive—quite the opposite, in fact; it was an ultra-light and super-compact vehicle, intended to carry no more than six people, including the driver.

According to Bollée's records, the weight of *La Rapide* was barely over the ton. It was a short vehicle, with a very low-slung chassis set between the front and rear wheels and giving easy access to an equally low-set driving position. The bodywork consisted merely of a sheet-metal shield at the front and side panels following the general line of the chassis frame. Aft of the driven axle, a light wooden 'tonneau' was attached to the chassis to provide four face-to-face seats on lengthwise benches. Although set low, the seat for the driver and his companion were above the engine, like that of *La Nouvelle*. Initially, *La Rapide*'s two cylinders were laterally mounted outside the chassis, with a direct drive through curved connecting-rods, not of course to the wheels themselves but to two shafts enclosed by a tubular axle which drove the planet-bevel yoke of a differential, which then acted on the tubular half-shafts carrying the wheels. Obviously this made for the maximum of compactness, but the direct drive to the wheels proved to be a mistake; Bollée resited his engine inside the chassis, driving the differential (still forming part of the back axle) through a chain, and thus with no intermediate gears. It will be remembered that this incorporation of the differential in the axle, used by Bollée in 1881, was credited to Louis Renault twenty years later.

The *Rapide* naturally kept the front wheel layout of *La Mancelle*, but differed in having grooved rims to the wheels. Bollée had, in fact, intended to put solid rubber tyres on this vehicle, which was designed to travel even faster than his previous carriages; but rubber of adequate quality was still not available and, after disappointing trials, he had to come back to the ring of hemp rope, which was to continue in use for some time. *La Rapide* No 1 reached 33mph, and a second version ordered by a M Leproust even attained 37mph, sixteen years ahead of any other mechanical road vehicle. Indeed, with the rubber tyre as yet unperfected, performance like this was also ahead of its time; it seems a reasonable presumption that Amédée Bollée's realisation of how remote his objective still remained was one of the secondary reasons why he now decided to call a halt to his work.

The main reason was the dashing of his hopes. From the end of 1880 Ernest-Sylvain Bollée stopped the loans to his son, who was heading for catastrophe; but being a fond father, he divided his three businesses between his three sons, on 1 January 1881, so that by running the bell foundry Amédée would be able to recoup the money dissipated in the building of automobiles. For two years and more Amédée and his family went short of everything in order to pay off his debts. But though times were so hard for him that he could not keep up the yearly payments on his patents, he still cherished the hope of one day being rewarded for all his efforts by the royalties owed to him by the Wöhlert locomotive works for the vehicles it was building on the basis of his plans. But in August 1883 the firm of Wöhlert collapsed, without having paid him a penny royalty on any of the twenty-two public-service vehicles it had built. Faced with the ruins of his last illusions, Amédée Bollée looked back on his life and cursed the past ten years which had jeopardised the very livelihood of his family; from that moment on, all he wanted to do was to follow in his father's footsteps in the bell foundry (which, in any case, he was later to raise to a high peak of renown). When, in 1885, a very rich nobleman in the area, the Marquis de Broc, placed an order for a successor to *La Nouvelle* fitted with an extravagant sixteen-seater, double mail-coach body, Amédée entrusted the work to his elder son (then just eighteen years of age).

In December 1886 the reply which Amédée Bollée gave to a request to supply a steam omnibus was something of a testament, and is worth quoting it in full: 'Since 1873', he wrote, 'I have been constantly concerned with steam vehicles, and have devoted to them considerable time and large sums of money, even much too large sums. I have, it is true, derived great satisfaction from the results which seemed, when they were first put to use, to be quite decisive; then with prolonged use came great disappointments. In the end I have come to wonder whether, with the present state of my practical knowledge, I should recommend the use of steam engines . . . .'.

This, in its moving terseness, is the judgement passed on his own work by the tireless pioneer who had brought the century-old dream of a steam carriage into everyday reality. Taking an objective and lucid view, he had realised that the steam carriage would never be the real answer to road locomotion. Though he had himself proved that it would work, and even reach high speeds, he had confirmed that the satisfaction of owning a steam carriage could belong only to a born mechanic; driving a steam car was an art, and taking it on the roads was an adventure.

But what is really astounding is that in that year of 1886, when nothing had yet been done to adapt the internal-combustion engine for locomotive purposes and when Bollée cannot have known what was afoot in Germany at the time, he should have stated clearly that 'the future belongs to the explosion engine, working on petrol fuel'.

No one could have foreseen the future more clearly: so how could the future fail to belong to this far-sighted man who had coped with the problem, who was the same age as Maybach, younger than Levassor and Benz, a whole ten years younger than Daimler and who already had thirteen years of solid personal experience in locomotive problems behind him? From our present vantage point we can see that Amédée Bollée was, without a doubt, the one person above all others who would have been able to bring about the birth of the motorcar. But his spirit of initiative had been broken by the years of drama, and he could do no more; he let his arms fall to his sides, and was content to encourage his sons to travel what he knew to be the right path. When their time came, however, it was to be already too late.

The character of Amédée *fils* is summed up in a few lines, written soon after his death in 1926:

'He was not concerned with advertising himself. As one of the most original of designers, his attitude to things mechanical was that of an artist who was permanently dissatisfied with his work. His whole objective was to build more and more sophisticated cars, better and better, for a hand-picked clientele.'

What, in the last analysis, prevented him from being one of the great pioneers of the motorcar? Ten years on his age: or else the ten extra years that his father could have devoted to the self-propelled vehicle which he had already brought so far towards fruition.

Let us go back to the 1886, when the lad who was to be given the mission of continuing his father's work and exploring the unknown world of the petrol engine, was nineteen years old. He had, it is true, a gifted father to take after; and by this age he had already shown considerable talent. At fourteen he had, with his own hands, built a miniature railway used by his grandfather to carry guests round his estate at Bel-Air. At eighteen he had built himself a pretty little two-seater steam carriage weighing 12½ cwt, with a combined chassis and bodywork, which was described by the journal *La Nature* and was the subject of an enquiry from Armand Peugeot.

But conquering the internal-combustion engine called for more maturity than one can muster at his age; and while the attention of Amédée *père* was occupied with the bell-making needed to restore his fortunes, he let his young son take the wrong turning. So it was that, in 1887, the first petrol engine produced by Amédée *fils* came to have a fixed piston and a moving cylinder to drive the crank-arms! And so it was again that the young inventor tackled a complex project—an airship engine in a nacelle, with a variable-pitch propellor—while the engine pure and simple had yet to be born. True that this rotary, three-cylinder engine (with a theoretical power output of 25hp!) was working by 1889, but it produced so much heat that it would have been impracticable, even if there had not been an inflammable balloon in its immediate proximity.

The following year saw a return to more realistic projects, and Amédée *fils* built a 6hp gas engine for his father's workshop and made preparations (which he did not follow through) for converting his steam carriage to petrol, drawing up plans in quick succession for two engines—both four-strokes, both horizontal, and with one and two cylinders respectively.

He was beginning to show, in this, his classic engineering technique of the years to come. Meanwhile, however, the young designer launched out into other bold schemes: in 1894 he invented no less than a gas turbine which, though it worked, had no practical future at all; and in 1895 he believed (like so many others before him) that he had invented the two-stroke.

However, he really had invented one thing—supercharging; for it was a fan which forced the gas mixture into the cylinder (or, alternatively, extracted it from the cylinder on the exhaust side). This was, in short, the turbo-compressor, the invention of which was to be ascribed to Rateau in 1917; but in that period of stumbling and groping the idea did not make any sort of sense.

It was this same year, 1895, that brought about a great change in the thinking of Amédée Bollée *fils* when, with his father and brother, he took part in the Paris–Bordeaux–Paris race on *La Nouvelle* (which by then was looking quite an old-stager). The lone inventor from Le Mans discovered what others were doing, and where they had got to. This was all that was needed to start him on his work: but he had lost eight years of practical experience in the meantime and was never able to make up for them enough to figure fully in the story of the birth of the motorcar. His ideas were to be outstanding, but success was to elude him.

The first car produced by Amédée *fils* was patented on 27 January 1896. The design bristles with *avant-garde* solutions, beginning with a metal chassis in the form of a rectangular frame mounted on long springs of unusual flexibility. Engine and transmission were located beneath the chassis, with the twofold intention of lowering the centre of gravity and leaving the top surface of the chassis completely free, so that it could be fitted with any sort of bodywork without restriction. Unfortunately, the first effect of this layout was to give the vehicle an old-fashioned appearance.

The engine, at the front, was of course horizontal; it followed the basic anatomy of the twin-cylinder engine of 1890, but in an up-to-date form. Not only were the two parallel cylinders and their water-jacket cast in a single block, with an enclosed crankcase, but the combustion chambers were hemispherical—and thus fifteen years ahead of the general state of technology. Better still, the piston crowns were concave, so as to achieve as far as possible a spherical combustion space.

This engine ran at 600rpm and, weighing 330lb, developed 6bhp. It was cooled by vaporisation and, in contrast to other engines, had a speed regulator acting on the exhaust. During the next six months it was to be fitted with an automatic carburettor with submerged jet, the first of its kind and seven years before that of Krebs and Claudel (See drawing opposite.)

The transmission was antiquated and progressive at one and the same time. Amédée *fils* had placed his gears at the back of the car. Motion was transmitted from the transverse engine to the mainshaft of the gear mechanism by long belts working under the best possible conditions—at constant speed and linking pulleys of equal diameter. Clutch action was obtained by sliding the belts sideways from fixed pulleys on to other, freewheeling pulleys.

From this point on, however, the transmission was chainless (well before Louis Renault). Following the layshaft carrying the differential, two longitudinal shafts drove each of the wheels independently through a bevel gear. This, then, was the system which was to arouse such interest when it reappeared in 1932 on a Grand Prix Alfa Romeo. Besides this, however, there was reaction and thrust acting through the springs (eight years before Hotchkiss), and bevel gears with spiral-cut teeth (seventeen years before Packard).

And even that was not all, for the Amédée Bollée car was also remarkable for its gearchange, clutch and brake controls mounted on the steering-column, its divided track-rod—thirty years before Panhard—and hardened ball-joints. Nor must we forget—thirty-five years ahead of all the others—a first form of central lubrication.

This was the vehicle which was to take part in the Paris–Marseilles race, to be shown at the 'Salon du Cycle' and built under licence by de

Dietrich; in the end, however, it was beaten by the fashion for a vertical engine and a common transmission.

In 1898 Amédée Bollée *fils* introduced in his engine a sophisticated improvement which seems to have attracted very little attention from the car drivers of the time. He replaced his original speed regulator with a progressive system which varied the lift of the exhaust valves, thus providing variable expansion with a constant compression. His next move, however, was spectacular in the extreme.

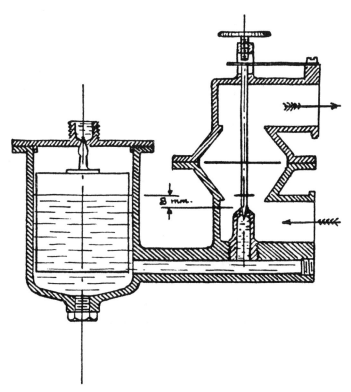

Despite still popular belief, the automatic carburettor
with a submerged jet was first designed in 1896 by
Amédée Bollée *fils*

For Amédée Bollée *fils* was the very first to appreciate the fact that, since the car was a moving body forcing its way through the resistance offered by the air, it needed to be given a shape ensuring good penetration. The Paris–Amsterdam–Paris race was the occasion for the first appearance of Bollée cars with what the public liked to call 'torpedo-boat' bodywork.

This bodywork, seen in plan, was pointed at both front and rear; thus it not only offered pointed bows which would pierce the air, but allowed the air to slide along its sides and merge smoothly again behind its stern. In elevation, the top and bottom of the body followed unbroken, horizontal lines. The idea came as a revelation; it was grasped at once, and motoring circles discussed endlessly what was known at the time as 'air lines'. Many observers, however, did not believe that the shape given to the rear of the body served any purpose.

Besides their futuristic design, the Bollée 'torpedo-boats' were also remarkable for being built in aluminium. This idea became fashionable at once, and within a matter of weeks the de Dietrich factories had orders for the Bollée 'torpedo-boats' amounting to more than a million gold francs. And one of these, made for the Comte de Paiva with coachwork by Rheims & Auscher, was unique in having the first sloping windscreen in the history of the motorcar.

The first torpedo-boat car was just as much a revelation for its creator, who found that the speed of his vehicle increased in a single leap to more than 37mph. He sensed that he was on the right road, and continued to follow it. After a brief diversion to design, create and develop the first band brake acting in both directions (which seemed miraculous to the

Amédée Bollée *fils* was the true pioneer of aerodynamic bodywork. This 'torpedo-boat' was built in 1899 for the Comte de Paiva. Note, in particular, the first sloping windscreen, thirty years ahead of Pinin Farina

drivers of the time), Amédée Bollée *fils* devoted himself to producing a 'super-torpedo-boat' of a car for the Tour de France in 1899. (See drawings below.)

The basic shape of the body having proved itself, he wanted to lower it as well—and did so in no half-hearted way! What he decided, in fact, a full two years before Daniel and fifteen years before Stabilia, was to *hang* the chassis *below* the rear axle. The engine moved from the front to the back of the vehicle, simplifying the transmission, but remained horizontal and was likewise placed below, and straddled by, the rear axle.

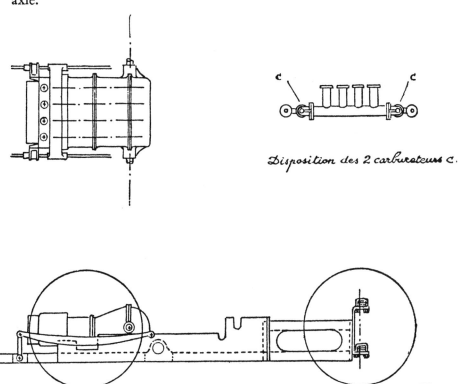

Disposition des 2 carburateurs C.

A remarkable anticipation of the future: the second 'torpedo-boat' produced by Amédée Bollée *fils* in 1899, with its low-slung chassis running beneath the axle and the rear-mounted, four-cylinder-in-line, twin-carburettor engine, the topmost point of which is barely higher than the wheel-centres

The chassis itself is extremely interesting. Its longitudinal members were made from U-section, strengthened with a flat strip of metal welded across the open side. Parts of the bodywork, in particular the 'bows', were made up from box sections which strengthened the chassis by acting as bulkheads. The engine, too, was sensational, since it had four cylinders cast in a single block with hemispherical combustion chambers, with two automatic carburettors (one at each end), and fitted with two starting handles so that the work of turning the engine over could be shared.

The flat, aluminium-clad surface on the top of the bodywork was matched by another flat surface of waxed canvas stretched across the underside of the chassis and giving a ground clearance (extraordinary at that period) of only 7½ inches. The shape of this vehicle, in fact, foreshadowed that of the present-day racing car, and was twenty-seven years in advance of any similar design. All these innovations paid off in performance, and during its trials the 20hp of the '1899 Torpedo-boat' gave it a speed of over 55mph.

Unhappily, the race itself ended in disaster for Bollée who, through shortage of funds, had been considerably delayed in his preparations and thus had to bring cars to the starting-line scarcely tested. This bitter experience brought home to him the pitfalls in racing, and he at once gave up competition motoring despite pressure from his licensee, de Dietrich.

From that moment on Amédée Bollée *fils* limited his interest to custom-built cars, striving constantly to improve their quality and refinement. In 1906 he launched his Type D, with a 10hp twin-cylinder engine, notable for its greater simplicity and substantial use of aluminium, adopted not for lightness but for extra strength at the same weight as its competitors.

The main feature of the Type D was a systematic attempt to achieve a stiff chassis, twenty-eight years before the principle was developed by de Ram. To obtain this stiffness, the chassis of U-section side-members had no less than five cross-braces, plus the exhaust silencer which was also fitted transversally. The vaporisation cooling system of earlier models had by now been supplemented by two condensers, housed one on each side of the bonnet, so that the cooling worked on the closed-circuit system.

One feature had been dropped from the Model D—transmission by means of two independent shafts, which was considered out of date by buyers of the day. In adopting chain drive, however, Bollée did not use tensioning rods; tension was obtained by varying the oblique angle of the link, the springs thus providing thrust and reaction just as before. Finally, the Model D was marked by its non-reversible steering (once again, the first of its kind), patented on 9 November 1900.

In 1902 Amédée Bollée *fils* continued his quest for perfect balance in a parallel twin-cylinder engine; his short-lived solution to the problem was a third piston acting counter to the other two and running in an open cylinder. In 1903 he turned his attention, successfully, to fuel injection, which was not new in itself but so far had not given satisfactory results; yet the motoring public, and the customers, did not rise to the bait. And in 1905 he designed the first automatic lubrication, on the splash system, which was the subject of a patent dated 31 October of that year.

This automatic lubrication system represented the first step towards the Type E design—a 30hp vehicle of which the first example took to the road in 1907. Once again, it bristled with new features. Up to then, valves had been in valve-chests arranged on both sides of the cylinder; the twofold bulk of the water-jacket and the valve spring meant placing the latter well away from the cylinder, resulting in a valve-chest of considerable size. Bollée's new idea was to close each valve by means of a tension spring located at the bottom of the pushrod, ie, beneath the water-jacket. In this way he managed not only to bring the edge of the valve to within 3mm of the cylinder bore, but also to house both valves side-by-side in the same valve-chest.

In the engine and transmission of the Type E there were ball-bearings and thrust-bearings in profusion. All the gears were fitted with an anti-vibration device consisting of a circular block built up from small leather discs, soaked in oil. The chassis was fitted with servo-brakes, and with belt-type dampers which allowed the axles to rise freely but dampened their rebound movement. The driver's position had been based on a statistical study of the measurements and preferences of a hundred car-owners, the arithmetical average being taken for each dimension. The gear lever disappeared, to be replaced by a second wheel concentric with and above the steering-wheel, and operated by one finger.

The clutch? This required no effort whatsoever from the driver, since its operation was independent of the spring tension; in his system derived from the steelyard balance scales (where the weight is moved along a long arm) Amédée fils noted that 'the pounds never meet the feet, so that any problem of foot-pounds is avoided'.

The Type E was also marked by a new automatic carburettor, patented in July 1907, and by the fitting of 'decompression cams' on the camshaft to make starting easier.

Hardly had the first examples left the works than Amédée Bollée fils began to design and produce fresh refinements. To start with, in order to quieten the engine, he tried to deal with the irregular speed of rotation of the camshaft, which met a variable resistance as it turned. The problem was solved by fitting a compensator whose variable resistance, added to the variable resistance encountered by the shaft, made up a constant total.

After dealing with the camshaft he turned to the magneto, whose successive phases (drive and resistance) were causing chattering of the driving dogs; this trouble was overcome by using another compensator. As an ultimate refinement, the springs were selected only after the body-work had been decided on and the resulting weight distribution was known.

Not surprisingly, perhaps, the price of the Type E in bare chassis form was 20,500 francs, at a time when the Rolls-Royce was selling at 21,000 francs.

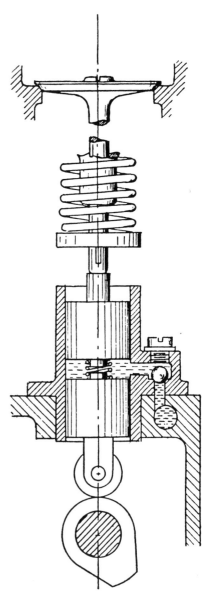

In 1910 Amédée Bollée *fils* invented hydraulic pushrods for the valves and fitted them to his production cars. His patent was taken up very much later by General Motors

At the end of 1909, Amédée Bollée *fils* took his Type E back to the drawing-board. He fitted it with combined ball-bearings and thrust-bearings of his own invention (patented on 27 October 1909), a flexible quadrilateral link between the rear axle and the chassis designed to keep the chain tension constant as the suspension worked, plate-type universal-joints (patented on 22 December), and a back axle allowing the camber of the wheels to be the same as it was when transmission was through two independent longitudinal shafts.

The most remarkable innovation, however, was patented on 23 April 1910—hydraulic pushrods providing an automatic take-up of valve clearance. (See drawing opposite.) Exactly twenty-three years later this Bollée patent was to reappear on the 1933 Cadillacs and Packards.

The four-litre Type F came on the scene in 1913, with an hydraulic torque damper fitted to the crankshaft. And this was to prove the swan-song of Amédée Bollée *fils*, designer, for the 1914–18 war put an abrupt end to the small-scale, manual production of twenty or so chassis a year, to which he had been limited for lack of the financial means to do any-thing else.

In 1919 the last Type F cars, for which the components had been in stock since before 1914, were assembled and sold, and the last page of the history of the *marque* was turned. From that time on, Amédée Bollée *fils* switched his works to making profit from a heat-forming process for piston rings which he had evolved in 1912; Amédée Bollée piston rings are still being produced today, in 1970.

There remains the story of the third Bollée, Léon, the younger brother of Amédée and the one whose name is best known. Contrary to what one might think, this story will not take long in the telling; for while he had the same gift for invention as his father and brother, his cast of mind was very different indeed.

The first thing we know about Léon Bollée, born in 1870, is his early passion for the bicycle, and the cycles fitted with floats which were the first inventions produced during his adolescent years. At the age of seventeen, he felt a vocation for bell-founding, and the start of his ap-prenticeship happened to coincide with his father's decision to undertake a complete revision of his tables of technical data. This involved a great deal of tedious figure work, and was all that was needed to encourage young Léon to take an interest in the machine; he revolutionised its prin-ciples, incorporating for the first time the facility of direct multiplication.

As a result he was able, at the age of nineteen, to present at the 1889 Exhibition a fantastic calculating machine containing more than 3,000 parts, which provided multiplication of one 10 digit number by another, or division of a 20 digit number by a 10 digit number, with a mere ten turns of the crank-handle! This machine brought him a visit from Thomas Edison, the gold medal of the Académie des Sciences, and a decoration from the Ministry of Public Instruction.

For the next six years, Léon Bollée devoted the whole of his time to designing automatic machines of all kinds: cash-registers, ticket-issuing machines, dating-and-numbering machines, arrival indicators for the lofts of pigeon-fanciers, machines for rolling cigarettes and even a machine for teaching people to swim!

This design work was so expensive for his father that the latter expressed great relief when, after returning from the Paris–Bordeaux–Paris Race, Léon followed his brother Amédée in declaring himself totally won over by the horseless carriage. The immediate result was the building of the well-known tandem tricycle, produced in order to get a head start on De Dion, whose orthodox tricycle was still very little known.

The remarkable aspect of this vehicle was the way it fitted perfectly into a period when the cycling fraternity were providing the bulk of the customers for the new and burgeoning world of motorised transport. Into an excellent framework was fitted the equally excellent horizontal engine produced by Amédée *fils*, together with a highly-ingenious combined control for the brake, clutch and gear-change, all working through a single lever operated with the left hand.

The Bollée tricycle was the subject of a vast publicity campaign launched by the *Petit Journal*, to whose editor, Pierre Giffard, the young inventor had given a share in his interests. The manufacturing rights were sold to Lawson, in Britain, for £20,000, while the French rights went to a company called the Société des Voiturettes.

In 1899 Léon Bollée designed a new vehicle—this time a four-wheeler and four-seater, still powered by his brother's engine, and with five gears provided by a single belt running round two stepped pulleys. The manufacturing rights for this machine were sold to Darracq.

In 1900, having made a great deal of money from his licence agreements, Léon Bollée suddenly changed course and decided to go into the business of building motorcars himself. He began by modifying the Darracq-type quadricycle, replacing its tubular, bicycle-style frame with a backbone chassis consisting of a single, tubular central beam of large diameter; this design was to be taken up again some thirty years later by German and Czechoslovak engineers. (See drawing opposite.)

From 1902, however, it was the introduction of the Mercedes cars which gave him his real aim in life; 'I shall', he declared, 'become the French Mercedes!' And in attempting to do so, he swept aside agreements made with his father and brother, under which he was to concentrate on light carriages while leaving the heavier cars to Amédée *fils*.

Although he did patent a large number of small detail designs (only one of these, the three-point engine mounting, is worth mentioning here), Léon, the young industrialist, proved content to follow the fashion in designing his cars, and during the remaining ten years of his life he built luxury models which the snobs of the world, captured by the lure of a skilfully-fostered fashion, fell over each other to acquire.

Between the famous three-wheeler *voiturette*, with two seats one behind the other, and the large cars which were to make him a fashionable manufacturer Leon Bollée produced this graceful *voiturette* in 1910. It featured independent front suspension and a rear-mounted engine and, thirty years ahead of its time, the idea of the backbone chassis based on a central tube

2 FREDERICK LANCHESTER

by ANTHONY BIRD

FREDERICK LANCHESTER (1868–1946). English:
Pioneer designer-constructor from 1895, as renowned for academic contributions to scientific
and engineering knowledge as for the brilliant but
unconventional cars bearing his name.

FREDERICK LANCHESTER

IT IS NOT easy to fit Dr Frederick Lanchester into a category. Because so many of his ideas were thought impracticable (being ahead of their time), and his last years spent in sadly straitened circumstances, he could be written off as a brilliant failure; because his contributions to knowledge range from colour photography to sound reproduction, from aerodynamics to optics, from motorcars to the musical scale, from relativity to a new theory of dimensions, from jet propulsion to poetry, he could be dismissed as a jack-of-all-trades, and because he delighted in deflating pomposity and let it be seen that he did not suffer fools gladly he might be called arrogant; yet his intellectual stature was such that he could not have been ignored in any age and it is probably not going too far to call him a genius.

Frederick William Lanchester was born in London on 23 October 1868, but soon after his birth the family moved to Brighton where his father was appointed architect and surveyor to a building estate. Though respectable, this job was not particularly well paid and the five Lanchester children could not be educated very lavishly. Frederick went to a small private school in Brighton where he was considered dull—mainly, according to his brother George, because of his manifest dislike of organised games. At the age of fourteen he went to the engineering school at the Hartley Institute in Southampton, and two years later won a scholarship to the Normal School of Science, now the Royal College of Science, in Kensington. Here he studied under Professors Goodeve and Vernon Boys, the latter an authority on gas engines, and also learnt workshop practice at night classes at the Finsbury Technical School.

By 1888 he had to support himself and started work at 6d an hour for a hack-draughtsman who provided drawings to accompany patent applications. After a few weeks at this rate, young Lanchester became self-employed and took the work on by contract, thereby putting his income up to £2 a week. He also patented his first invention; this was for an instrument called an 'isometrograph' which took the donkey-work out of shading and hatching on engineering drawings.

Later in the same year he took a job with T. B. Barker, sole proprietor of the 'Forward' Gas Engine Company of Saltley, near Birmingham. The wage of £1 weekly was low even by 1888 standards for a trained engineer, but there seemed to be good prospects. A clause in the draft service agreement provided that any improvement Lanchester might make should become the property of the company—of Barker himself in fact. Young Lanchester objected to this and Barker struck it out with the sneering remark that it was of no consequence as Lanchester was unlikely to be able to teach him anything. Within six months Frederick had patented his pendulum inertia governor, and Barker was paying him a 10s royalty on each engine fitted with it, and within a year it was obvious that Lanchester was the best man to step into the shoes of Charles Linford, works manager and designer, who died from tuberculosis. Linford left no proper drawings and had recorded all the specifications in his private notebooks, and in making good these deficiencies Lanchester took the opportunity to re-design and modernise the company's range of engines.

Soon after he had taken charge of the works Frederick sent for his fifteen-year-old brother George, and had him apprenticed to the firm, paying the premium from his own pocket. It was the start of a long collaboration and George proved so apt a pupil that he was able to take over his brother's job in 1893; as soon, that is, as his apprenticeship was over and while still in his 'teens. Competition in the gas-engine business was far too keen for mere nepotism to have dictated this appointment; there must have been merit as well.

The year before Frederick relinquished his post in favour of his brother he had rented a small shed from his employer and set up his own experimental workshop and laboratory. Here he made his first small petrol engines (with the economical and ingenious wick vapouriser which was used on all Lanchester cars until 1914), one of which was used to generate light for part of the factory and another which powered the first English motorboat. He also developed a flame impulse starter for gas engines, which solved a real problem in the days when even the biggest engines had to be started by the risky procedure of hauling on the rim or spokes of the flywheel.

Lanchester's starter was patented in 1890, and the application was handled for him by Dugald Clerk, who was not only an expert in patent law but also an authority on internal-combustion engines. In 1891 Clerk patented a somewhat similar starter, which relied on a single high-pressure impulse instead of Lanchester's series of low-pressure impulses, and there was for a time some dispute about priority between the young inventor and the established expert. This difficulty was resolved, and Clerk granted Lanchester a half share in his form of the device, which was sold as the Clerk-Lanchester Starter and yielded a fair income.

Towards the end of 1892 Lanchester went to the United States to try to interest gas-engine makers there in the starter, but the engine business in the States was relatively backward, and the engines being made there were mostly small enough to start by hand without danger.

On his return from America, Lanchester severed his connection with Barker and devoted all his time to experimental work. To augment his income from royalties, he set up a small business to make bicycle pedals and hubs, with damp- and dirt-proof ball-bearings which he had invented: he persuaded his second brother, Frank, to leave his job in a London bank to manage this venture, and Frank subsequently became sales manager of the Lanchester Motor Company and a leading figure in the motor trade.

Most of Frederick's time at this period was given to experiments and calculations in connection with heavier-than-air flying-machines. He made innumerable models of gliders and elastic-powered aeroplanes, and postulated his theory of vortex sustentation which was subsequently established as a fundamental part of aerodynamic knowledge under the name of the Lanchester-Prandtl theory.

Having established the possibility of securing inherent stability in his powered models, Lanchester was anxious to go on to full-scale machines. He considered Lilienthal's gliding experiments valuable but inconclusive, and early in 1894 he approached Dugald Clerk to ask for his help in financing the development of an engine light enough for an aeroplane. This, it must be remembered, was ten years before the Wright brothers flew and Clerk's answer was understandable though disappointing. 'Larnchester,' he said, 'you may be right and I am sure you could do it. But you must consider your reputation. If you were to put forward such a proposition to business men now your reputation as a sane engineer would be ruined.'

This advice led Lanchester to turn his attention to motorcars until the time should be ripe for the aeroplane. After studying French and German developments (of which he thought poorly) Fred set about designing a motorcar from first principles with none of the usual borrowing from established carriage or cycle trade practices. Helped by George, he began construction in 1895 and the finished article, Britain's first four-wheeled petrol car of wholly native design, made its first trial run before dawn on a February morning in 1896. This was nine months before the 'Emancipation' Act took effect, and at a time when 'road locomotives' were still subject to the 4mph limit which the Lanchester brothers coldly ignored.

One of the characteristics of early motorcars which Lanchester disliked was the vibration from their large (but feeble) unbalanced single- or twin-cylinder engines. This he considered destructive to the mechanism and unacceptable to the potential customer, but the obvious solution of using a larger number of smaller cylinders, with lighter reciprocating parts, was hampered by the manufacturing problems involved in making long, multi-throw crankshafts. Therefore, the first Lanchester car engine, though it had but one cylinder (of 4¾in bore and 4½in stroke), was much less offensive than most as the solitary piston gave motion to two connecting rods, side by side on the gudgeon pin, and thus to two cranks and flywheels. The cranks were geared together and rotated in opposite directions and most of the forces were balanced by reverse rotation. The engine had a mechanical inlet valve, the patented wick carburettor and low-tension ignition from batteries; it drove the car through a two-speed epicyclic gear and chain to the live axle which contained an ingenious but simple overdrive mechanism.

The car ran well but was underpowered (or overgeared), and within a short while it was fitted with an 8/10hp two-cylinder engine which had been designed alongside the earlier one. The new engine was also balanced by reverse rotation, with the two cylinders horizontally opposed and the pistons coupled to the counter-rotating cranks by a 'lazy-tongs' parallelogram of six connecting rods. The centre of gravity of the parallelogram, which had true harmonic motion, was always at a point half-way between the crankpins and was consequently neutralised by the centre of gravity of the reverse-rotating crank balance-weights. The counter-rotation of the cranks and flywheels also ensured that the inertia forces were balanced within the engine and were not communicated to the car frame.

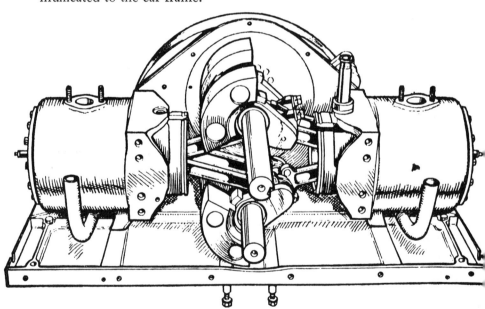

The Lanchester twin-cylinder engine with twin balanced crankshafts each carrying three connecting-rods

Page 61   Dr Frederick Lanchester, circa 1930

Page 62 (*above*) The first Lanchester car of 1895, with a two-cylinder engine
fitted in 1897 and experimental wheel steering added in 1899, photographed in
1931; (*below*) Mr George Lanchester on the 1897 8hp two-cylinder Lanchester
at Camberley, 1958

Page 63 (above) The 20hp four-cylinder Lanchester tourer, 1904/5; (centre) the 20hp four-cylinder Lanchester engine and gearbox assembly circa 1905, (prototype 1904); (below) the 20hp four-cylinder Lanchester chassis, 1905

Page 64 (above) A 1908 four-cylinder landaulette which remained in everyday use in Tunbridge Wells until 1939. Photograph taken the day the car was delivered new in 1908; (centre) 1913 38hp six-cylinder torpedo tourer. The owner, F. W. Hutton-Stott, at the wheel; (below) 1915 production Lanchester Sporting Forty designed by George Lanchester. This car has a side-valve engine

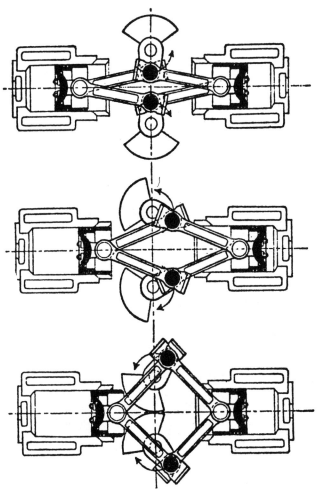

Three phases in the movements of pistons and connecting-rods in the Lanchester engine: (*top*) the rods
nearest to horizontal centre line of the opposed cylinders, lift piston on power stroke; (*centre*) cranks at
TDC for right piston, which is at end of compression
stroke; (*bottom*) the left piston is on BDC and the rods
are approaching their widest relative angle

This remarkable piece of ingenuity formed the basis for the production car engines of 1900–5 which differed only in detail and cylinder
dimensions and, after 1902, in being made available with water cooling.
For the original type was air-cooled by ducted forced draught. Other
notable features were fully automatic lubrication, a flywheel magneto
to supply current for the patented low-tension 'igniters', which could

be adjusted whilst the engine was running, and a singularly clever form of valve gear which utilised one large valve in each cylinder, and a mechanical disc valve which alternated with the main valve in communication with exhaust or inlet tracts as required, thereby keeping it cool and preventing many of the valve troubles prevalent in the days before heat-resisting steels were developed.

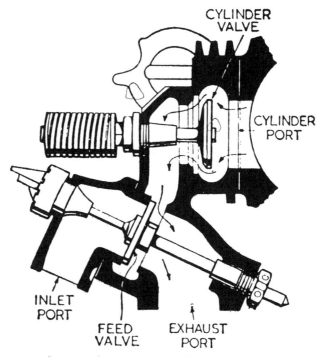

Lanchester valve gear: a single valve in the combustion chamber functions on both inlet and exhaust strokes in conjunction with a distribution, or 'feed' valve, which alternately admits mixture on the inlet stroke and allows escape of burnt gases on the exhaust stroke

When fitting the new engine to the original car Lanchester also abolished the exposed chain transmission, which offended his mechanical sensibilities, and substituted shaft drive and his famous 'enveloping' or 'hour-glass' worm gear. This was based on the Hindley gear which had long been known as a textbook example of a perfect worm drive, but for which no manufacturing techniques had then been evolved. It was typical of Frederick Lanchester that he not only evolved a satisfactory method of making the gears, but designed the necessary hobbing machinery to do the job. It is also noteworthy that the Lanchester was the first car in the world to be expressly designed to run on pneumatic tyres.

A second car followed in 1897. It was driven by a similar two-cylinder engine but was smaller and lighter, and consequently performed better. On its first trial it covered sixty-eight miles at 26mph, and in 1899 Frederick Lanchester was awarded a special Gold Medal for 'excellence of design' at the Automobile Club's Richmond Trial. The 'Gold Medal Phaeton' subsequently covered the course of the Club's 1,000 Miles Trial in 1900 and is now preserved in the Science Museum in London.

Behind the construction and trials of these cars, and much more experimental work, were the efforts to find financial support. A small syndicate to exploit the Lanchester patents was formed late in 1898, and was floated as the Lanchester Engine Co Ltd in December 1899. Before the company was formed, the syndicate authorised the construction of a third car, mechanically identical with the second, on which the backers insisted that the Lanchester brothers should fit a coachbuilt body as they considered Fred's neat, tubular-framed body on the Gold Medal car too unorthodox. This bodywork, built by a renowned Birmingham coach-builder, split in two during the 1,000 Miles Trial; the disaster seems to epitomise the lack of harmony between Lanchester and his directors.

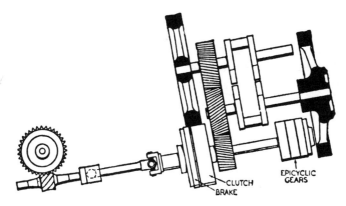

The twin-cylinder Lanchester engine was tilted back-wards in the frame. The diagram shows the two crank-shafts, helical gears driving the countershaft, clutch and brake, epicyclic gear drums and short shaft to the worm-drive back axle

The complete plan for the production model was laid down several months before the company was launched, and it was so perfect an example of designing a complete mechanical entity from first principles

that very few details had to be modified during the production life of
five years (the model continued in production alongside the new four-
cylinder car in 1905). One cannot describe the car in detail here, but it
was a remarkable piece of work with a composite steel and aluminium
chassis, integral with the lower part of the body, of exceptional torsional
and beam strength, mounted on long cantilever springs front and back
with parallel motion linkwork. It had three-speed epicyclic gears on
Lanchester's patent compound system (the basis of the modern 'automatic' gearbox) with pre-selector control, and a degree of riding comfort, road-holding and braking not generally common until some forty
years later. Though unlike any other car in practically every detail, the
production Lanchester Twin was no mere freak, but a sound, reliable
and handsome motorcar. The vibrationless, midships-mounted engine,
now enlarged to $5\frac{1}{4}$in × $5\frac{1}{16}$in ($4\frac{1}{4}$ litres capacity) was rated at 10hp and
gave the car a top speed of 40mph at an astonishingly economical
25/30mpg. Out of deference to public hostility to air cooling, a water-
cooled version (called the 12hp) was available optionally from 1902, but
to demonstrate faith in the air cooling, Lanchester also produced in
1903 an air-cooled 16hp model, with the cylinder bores enlarged to $5\frac{1}{2}$in,
which was capable of nearly 50mph. The 16hp air-cooled engine, despite the size of the pistons, could safely exceed 1,750rpm thanks to the
perfect balance. This was a very high-speed engine by 1903 standards,
and the output of 23bhp at peak speed was above average; the crankshafts ran in roller bearings.

The Lanchester was the first car to be designed entirely by one man,
owing nothing to any other source. Every detail of mechanism and
bodywork, even down to such items as fitted drawers for tools and
spares, the 'luggage deck' and a folding picnic table came from Frederick
Lanchester's hand and brain. Day-to-day organisation of production fell
to brother George who was also chief tester, apprentice-trainer, jig and
tool superintendent and layout man for such advertisements as the
struggling company could afford.

In the technical perfection of the car lay its commercial weakness.
Lanchester's directors (who were also the chief shareholders) naturally
wanted a return on their money, and continually urged their designer to
cut corners and produce cars by the knife-and-fork methods used by
most contemporary manufacturers. Lanchester looked further ahead;
his ideas embraced the whole concept of mass-production short of the
moving assembly line; he insisted on complete interchangeability of
machined components and in many instances had to design special tools
and evolve new processes to meet his needs. He considered, for example, that the Whitworth thread was too coarse in diameters below

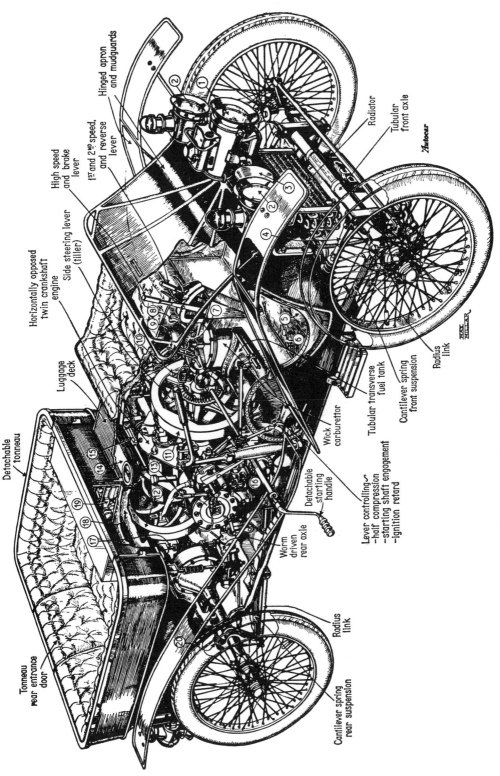

Hinged apron and mudguards

Radiator

Tubular front axle

Autocar

High speed and brake lever

1st and 2nd speed, and reverse lever

Horizontally opposed twin crankshaft engine

Side steering lever (tiller)

Luggage deck

Detachable tonneau

Tonneau rear entrance door

Cantilever spring rear suspension

Worm driven rear axle

Detachable starting handle

Lever controlling—
—half compression
—starting shaft engagement
—Ignition retard

Wick carburettor

Tubular transverse fuel tank

Cantilever spring front suspension

Radius link

Radius link

MAX MILLAR

The 1903 Lanchester, from a drawing by Max Millar

¾in, and consequently made his own taps and dies to produce his own Lanchester 'M' thread many years before the very similar BSF thread was established. He could find no manufacturer able to produce the roller bearings he had designed to sufficiently close limits, and consequently designed his own machinery for making them. He found many of the craftsmen he and George were training had difficulty in reading engineering drawings and so he evolved his 'unilateral' system of dimensioning. Even such simple things as 'go' and 'not go' gauges were not available commercially and had to be designed and made.

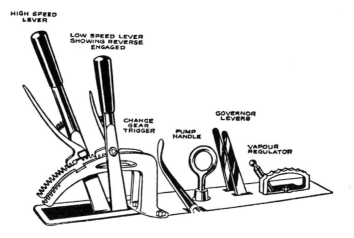

Controls of the twin-cylinder Lanchesters, which were
situated between the front seats

Other manufacturers went ahead by relying on the plentiful supply of cheap skilled labour which made hand-fitting of most components commercially acceptable for a time. Lanchester would have none of it, and so it was 1901 before the new Lanchester works at Armourer Mills, Birmingham were ready to produce cars for sale. Time has proved him right, but his directors were able to complain of lost ground and to point to competitors who had outdistanced them; they urged him to cut costs by abolishing some of the distinctive Lanchester features, like the worm gear, and could not understand his ideas. The delay in starting production had eaten into the reserves of an under-capitalised company, and also permitted the conventional, or 'Panhard' type of car with its vertical forward-mounted engine, to become established in the public mind as the divinely ordained 'correct' form.

Despite the mutual hostility developing between the engineering and financial sides of the business, the Lanchester company sold some 450 of the two-cylinder cars, which was no inconsiderable output for those early years. Trouble came in 1904 when a reasonable profit had been earned but more capital was needed to end the hand-to-mouth way the business was then conducted: as general manager, Fred Lanchester was often hard put to it to find the ready cash for the men's wages. Additional funds were also needed to finance production of the new four-cylinder 20hp car which was on the drawing-board. Instead of increasing their own investments, as they could have done, the directors launched a public issue of 42,000 £1 shares to bring the issued capital up to the authorised figure of £100,000; but most motor companies were in difficulties, their shares were unpopular on the Stock Exchange, the issue failed and the company passed into the Receiver's hands in March 1904.

During the year of bankruptcy, the very successful four-cylinder 20hp model was launched (it was followed by a 28hp six-cylinder in 1906), and a small profit was made under Frederick's management. The company was then reconstructed by the director/shareholders taking up the additional shares (as they might have done before bankruptcy), the shoe-string conduct of affairs was eased and a satisfactory return on capital was earned for the next twenty-five years. Unfortunately, Frederick Lanchester had no share in this prosperity. He was a poor man and his only shareholding had been allotted to him in consideration of his patents and equipment. With typical generosity he had given a tenth part of his holding to George and Frank and as a result of the re-organisation, being unable to put up any cash, he had had to surrender most of the remainder. He ceased to be general manager but remained as designer with his salary reduced from £450 to £250 pa. Though he thought the arrangement unjust, his desire for the company to prosper was so strong that he did not cavil at the unpleasing conditions of settlement, and he was made free to take outside work.

Despite the financial difficulties and directorial disagreements, there were plenty of lighter moments in these formative years, and Lanchester and his brothers had the satisfaction of knowing that their name was attached to a superb piece of engineering of unique design. Because of its unconventional appearance, the Lanchester car provoked some sales resistance, but it is significant that a very large proportion of those who bought a twin-cylinder model in the early days remained faithful Lanchester fanciers until the unhappy day when the company succumbed to the crisis of 1930-1, and was taken over by Daimler, who merely used the Lanchester name as a label for a new range of cut-price models with which they hoped to capture a slice of the middle-class market.

All pioneer motorists were constantly in trouble with the police and the Lanchester brothers were no exception. They were occasionally able to hoodwink the law by exchanging cars, or even identities. After the Richmond Trials of 1899 Frederick left it to George Lanchester and Archie Millarship (who became sales demonstrator) to take part in trials and rallies but he occasionally used local club events for road-testing cars before delivery. On one such occasion he changed cars with George, who had doubtless exceeded the speed limit earlier in the day, and contrived to abduct a policeman who jumped aboard the car as it moved off, demanding his name and address. This incident forms the central theme of Kipling's splendid story *Steam Tactics* and the 'Octopod' car in that story is Kipling's own 10hp Lanchester which was the sixteenth to be sold to a private customer. The car was, incidentally, delivered to Kipling at Rottingdean by George Lanchester himself, which sheds a revealing light on the scale of the motor business in 1901.

Between 1905 and 1909 Frederick was responsible for the constant improvement of production methods, and for detail modifications to the cars which included the change, optional at first, from his dynamically stable side-lever steering to wheel steering to suit, as he put it, 'the standardised chauffeur'. In 1909 he became consultant to the Daimler Company, but also served his youngest brother George as consultant in evolving the new 25hp and 38hp model Lanchesters out of the earlier types. This was in 1910-12, and thereafter Frederick had no connection with the pioneer British motor company which bore his name.

As consultant to Daimler (who were taken over by BSA in 1909) Frederick Lanchester did much to make the Knight double-sleeve-valve engine a commercial proposition, though its incurably smoky exhaust offended his susceptibilities. He also induced the company to use his worm gear final drive which, rather unfairly, they referred to in advertisements as the Daimler-Lanchester gear. As this period coincided with the short-lived association with AEC, the worm gearing was also adopted for London buses and soon became general for heavy lorries, despite the experts who had said it would never be of the slightest use except for light cars.

His most important contribution to Daimler solvency was the Lanchester crankshaft vibration damper produced to cure the torsional vibration which afflicted most early six-in-line engines. Daimlers had rashly embarked on a six-cylinder model, and had built a number which they dared not sell, so badly did they vibrate at critical speeds. In his six-cylinder Lanchester engine Frederick had avoided the difficulty by using a crankshaft of much greater stiffness and journal diameter than was then common, by placing the flywheel at the front to balance the rotating masses of the epicyclic gear drums at the back and by using an oversquare bore/stroke ratio; but the damper could easily be adapted to any engine and its use released £20,000 worth of otherwise unsaleable cars. The torsional vibrational damper was followed by his harmonic balancer for four-cylinder engines. These two inventions were widely used (the torsional damper still is) and brought him in a good income from royalties, and substantial sums from sale of manufacturing rights in America. With his first taste of reasonable prosperity, Lanchester was able to marry and to indulge his liking for good wine and his passion for yachting.

Lanchester's work during this period included the invention of the modern method of making piston rings. The Lanchester cars, incidentally, were fitted with oil-control rings at a time when most manufacturers were content to let their engines smoke. Other innovations, now commonplace, were the use of full-scale high-pressure lubrication, at 40psi in 1905, lightweight steel pistons, and splined shafts in place of the square or keyed shafts normally used at the turn of the century. Lanchester's four-cylinder and six-cylinder engines had horizontal overhead valves and were notably economical and efficient. The 38hp engine, of 1910 onwards, was of 4·8 litres capacity and developed 48hp at 1,400rpm and about 63hp at its optimum pace of 2,000rpm. This compares favourably with the 'Silver Ghost' Rolls-Royce engine for which almost identical horsepower figures are recorded from a capacity of just under 7½ litres.

Unfortunately much of Lanchester's work for Daimler was devoted to prestigious but unprofitable projects such as the Renard Road Train and the KPL (Knight, Pieper, Lanchester) petrol-electric bus, or the remarkably advanced motor rail-car for which Lanchester designed a seven-speed epicyclic gearbox controlled by electro-magnetic clutches. A good deal of his time was taken in preparing 'papers' on various automobile and allied subjects for the Institution of Automobile Engineers, or its older cousin the Mechanical Engineers Institution. Lanchester's contributions were always good value (his paper on Worm Gears and Worm Gear Mounting could be profitably studied by any production manager today), usually controversial and frequently tendentious. His large frame and imposing presence matched the vigour with which he joined in the debates which followed his own papers and those of others. He was once embroiled in heated argument with his great contemporary, Laurence Pomeroy, who contradicted his theory of the inertia limitation of piston speeds. At the time, Pomeroy appeared to win the argument but as engine speeds rose far above the limits practical in 1909 Lanchester was found to be right and some fifty years later Pomeroy's son, as

E

technical editor of *The Motor*, made handsome amends for his father's rejection of the theory.

Because Lanchester was a very large and impressive man of commanding presence with an exceptionally quick mind, he came to be regarded as rather 'grand' and insensitive to the feelings of others. This was not really so, as his happy marriage and his cordial relations with his brothers show; also many of his ideas on what a motorcar should be like were influenced by a fundamental understanding of the human animal.

In the early years of the century the Lanchester car was often decried by self-styled experts because of its unconventional appearance and unorthodox machinery, but nobody had anything but praise for its superb suspension, which gave easy riding to the occupants and long life to the tyres. The combination of rigid chassis and parallel motion linkwork was responsible for good handling, and the comfortable ride derived from the fact that Lanchester allowed a much greater range of spring movement than usual, and based the rate at which the springs vibrated on the normal walking pace of an adult. By doing this, he argued, the motion of the car would feel 'natural' and comfortable; he similarly arranged the front seat so that the driver's eye level should be roughly the same as that of an average man walking so that his judgement of speeds and distances should be normal and accurate.

When the time came to take advantage of improved techniques by producing cars with four-cylinder and six-cylinder engines, Frederick still showed his concern with the human factor. Various technical considerations decreed that the new engines should be vertical and placed conventionally, in the front of the chassis, but it offended Lanchester's sense of the fitness of things that the human cargo should be cramped in order to make room for a long bonnet. By making the engines exceptionally compact and narrow by the standards of the time, they could be disposed in a narrow casing between the dashboard and the front seat, and great ingenuity went into making all the vital parts easily accessible. The engine casing separated the leg-space of driver from that of front passenger in much the same way as the gearbox or transmission 'hump' of many modern cars.

On a wheelbase of modest length, which helped keep the chassis stiff, the 'Lanchester Engine Position' allowed the bodies to be given very wide doors and generous interior space. In 1908, for example, the 'short' 20hp gave 25 inches of space between the leading edge of the back seat and the backrest of the front one, on a wheelbase of only 9ft 5in, whilst the 'long' 28hp gave no less than 42 inches and allowed for yard-wide rear doors at a time when twenty inches or less was regarded as normal.

When commercial considerations obliged the directors to instruct George Lanchester to bring out the company's first 'conventional' car (the 'Sporting Forty' of 1914), the technical press had rather changed its tune since the days when Lanchester's designs had been suspected of being 'too clever'. *The Automobile Engineer* reporter wrote:

> The fact that a company such as the Lanchester have found it impossible to educate their potential customers to an appreciation of the correctness of Lanchester principles is a matter of regret. Considering . . . the first logical principle of design . . . there is only one form for an automobile to take, and this form has been characterised in Lanchester cars since the early days.

The 'Lanchester Engine Position' is now part of the ABC of commercial vehicle design, and though nobody now uses cantilevered leaf springs the suspension periodicity of most medium to large passenger cars is curiously similar to that used by Lanchester at the beginning of the century.

We are chiefly concerned here with Frederick Lanchester as a motor-car designer, but his aeronautical work cannot be wholly ignored. The years 1907 and 1908 saw the publication of his *Aerodynamics* and *Aero-denetics*, the two together constituting a thousand-page advanced work on *Aerial Flight*. The full importance of some of the theories mathematically treated in this work was not fully appreciated at the time, but the books aroused great interest at home and abroad. They involved the author in more controversy, and led to his writing numerous articles for technical journals and reading more 'papers' before learned societies. It is, incidentally, typical of Lancunian industry that he took singing lessons in order to improve his breath control for public speaking.

For a short time Lanchester was consultant to White and Thompson in their attempt to produce an aeroplane less frail and unstable than the string-and-sticks Wrights and Voisins of the day. The Lanchester-influenced aeroplane never flew, because of a disaster during take-off when the undercarriage collapsed after striking a partly-concealed boulder in the sands of their chosen runway on the foreshore near Bognor. The accident at least proved the strength of Lanchester's stressed-skin aluminium and steel-clad wings and fuselage. Soon after, Lanchester was appointed to the Government's Air Advisory Board and thought proper to resign from a firm which might tender for government contracts. Thompson's contribution to aviation was subsequently acknowledged by the Royal Commission on Awards to inventors and he, in turn, acknowledged his indebtedness to Lanchester's work.

Lanchester's writings were largely intended for experts, but in 1916 he published *Aircraft in War*, consisting of articles, with addenda, which had been written in 1913; at a time that is when the Service chiefs still said that the aeroplane could serve no purpose in war. This was a remarkable work of true prophecy; not the prophecy of Wellsian science fiction, in which the author may overcome technical hurdles by the convenient 'invention' of a hitherto unknown material or process, but an accurate forecast of events argued from deduction and logic and containing his now famous N-square-law. *Aircraft in War* reached a large public and played a part in influencing strategy.

Lanchester's other direct contribution to war effort lay in his work on the Advisory Board on Aeronautics, the Admiralty Board of Inventions and Research and the Air Inventions Committee. For Lanchester this was not just a matter of attending meetings, but included a great deal of original work, at Daimlers, the National Physical Laboratory and the Royal Aircraft Establishment to investigate problems and prepare reports. These, as always, provoked argument and sometimes hostility and provide the reason, no doubt, why Lanchester's war service was not recognised by any of those decorations which were distributed so lavishly. As his biographer, Dr Kingsford, points out, he undoubtedly saved the country a great deal of money but when he knew he was right, as he so often was, he could be severely outspoken to those who disagreed.

After the war, Lanchester's automobile work included a spell with Wolseley as consultant in designing and starting production of their post-war models. This and other consultancy jobs were in addition to Daimlers' demands on his time, but in 1926 he agreed to work for them exclusively in return for an increased annual fee.

An independent venture of the early twenties was concerned with a cyclecar intended to sell retail for £65 or less. George Lanchester helped with the design and saw to the construction of five prototypes. The first two had BSA V-twin engines, but the rest were powered by Frederick's own design of flat-twin engine which, with one of his jumps into the future, had a direct-injection fuel system instead of a carburettor. The cars were rear-engined and had elegant Lanchester-designed bodies of 'Consuta' or 'stitched wood' construction by Saunders of Cowes. The most bizarre originality was displayed in the chassis, suspension springs and front axle beam which were also of wood. Two forces combined to stop the cyclecar project; the first was Fred's passion for refinement which led him to design a clever but elaborate petrol electric transmission which would have made the 'Wooden Lanchester' simple to drive but too expensive; but the more important factor was the appearance of the Austin Seven in 1924 which put paid to many cyclecar ventures.

The unwieldy management structure of the BSA/Daimler Group

cannot be analysed here, but its shortcomings led to a lot of expensively unproductive design work for which the technical staff in general and Lanchester in particular were subsequently blamed. In 1925, a subsidiary company called Lanchester Laboratories Limited had been formed to exploit some of Frederick Lanchester's (Dr Lanchester now) inventions. Scarcely had the agreement been signed when the parent company had to borrow back the working capital advanced, and Lanchester Laboratories was unable to go ahead. The financial wind blew ever chillier and in 1927 Lanchester, in common with Daimler's own management staff, had to accept a cut in salary. Finally, as the depression swept through Wall Street and beyond, the board decided they could not afford the luxury of a full-time consultant and Lanchester's main source of income was abruptly stemmed.

Worse still, he was left holding the baby—Lanchester Laboratories Limited. He had either to accept a cash sum in consideration for some fifteen patents of inventions he had assigned to the subsidiary, in which case he knew that Daimlers would let them die, or he had to buy the Daimler shareholding. Rashly, he chose the latter course though it took all his capital. As financial panic swept well-established businesses (including the Lanchester Motor Company) off their feet, Frederick Lanchester took on the job, at the age of sixty-one, of setting up a manufacturing business with practically no working capital.

Buildings were erected on a small site he owned and production started on a line of loud-speakers and transformers. The principal product was Lanchester's patent 'Euterpephone', an acoustic-tube moving-coil speaker with a diffractophone aperture to give accurate distribution of tones of different pitch. It was a remarkably effective instrument of typically Lancunian ingenuity, but disaster followed on disaster.

Lanchester gave the job of production manager to a technical assistant who had been appointed by Daimlers when they originated the scheme; he thought this only fair as the man had moved house in order to take the job, but unfortunately he proved ineffective as a manager though admirable as a technician. Lanchester found him another job and took over production himself, but there was no money for publicity (indeed, there was almost no money for loud-speakers) and the business had to work on a mail-order basis. If Lanchester had closed the firm in 1932 he might have saved most of his capital, but he shrank from discharging workmen at the height of the slump. He struggled on until 1934 when his health gave way and Lanchester Laboratories had to close.

The remaining years found Lanchester being loaded with honours and distinctions by learned societies but too shaky from the onslaught of Parkinson's Disease to hold a pen, and growing blind from cataract; he was now recognised as the Grand Old Man of automobile engineering but was too poor to keep a car. He had little income beyond a minute pension from the Society of Motor Manufacturers and Traders.

As he could not use a pen, he learned to work a typewriter, when his sight made this impossible he dictated to his wife and the scientific papers and other writings continued. These included a volume of poems, under the pseudonym of Paul Netherton-Herries, and a book, *Relativity*, published in 1935. His scientific papers ranged over subjects such as 'Span' (in which he envisaged the plastic-pneumatic dome), radiation, jet-propulsion, airships, skin-friction, the musical scale and so forth, but between 1922 and 1939 he contributed fifteen important papers to the Association of Automobile Engineers. Amongst the most important were two long dissertations on independent suspension (written in collaboration with George, who carried out the practical experiments), and it is interesting to note that one of the many inventions for which he was unable to find a market was a system of suspension which was 'Hydro-lastic' in all but name.

He had been made an Honorary Doctor of Laws by Birmingham University in 1919, and a Fellow of the Royal Society in 1922. Other distinctions included the Royal Aeronautical Society's Gold Medal in 1926 and the Guggenheim Gold Medal in 1931. His last honour, and the one he most valued, was the Institution of Mechanical Engineers' James Watt Medal in 1945. 'Dr Fred' was by then too frail to attend the presentation and 'Mr George' deputised for him. He died in the following March.

Sir Harry Ricardo, who knew him well, wrote that he was:

> ... a great scientist, a great engineer, a mathematician, an inventor, a true artist ... and a poet and philosopher. ... His mind worked so quickly ... that few could keep pace with it, and this rendered him rather intolerant of their slower processes. ... Like all great inventors he had to endure the mortification of seeing his own inventions re-appear years later, perhaps in slightly modified form and under another name. This he took very hardly, for he failed to recognise and accept that this, inevitably, is the fate of all such.

This justly sums up Dr Lanchester's strength and weakness. To working men, students and apprentices he was gentle and generous, but he could be crushing to those of his own standing by failing to realise that it was given to few men to match his intellectual stature. This was really humility, but often appeared to be pride. He had a quick wit which was often caustic. After a particularly tedious meeting with his directors in the early days of the Lanchester Company he was heard to remark: 'Well, they seem to change their minds pretty often—but then, if I had a mind like any of theirs I'd change it as soon as I could.' During the discussion following his paper on Worm Gears, he was asked for his opinion of the Wrigley Company's product and replied that although the Wrigley Worm might be useful for fishing he could see no future for it in motorcar transmissions.

Lanchester's favourite relaxation was sailing and in his prosperous years he spent as much time as he could aboard his yacht; fortunately, his wife shared his tastes and as they had no children domestic ties did not intrude. In less prosperous days there was always music as a solace, and those who knew him say there is little doubt that Frederick's singing voice could easily have been trained to operatic standards. He liked good food and wine, and comfortable surroundings, but he cared little for his appearance, which was impressive but untidy, and he was always quite happy with cheese, pickles and beer if pressure of work or lack of money put the caviare and claret out of reach.

From a mere pen picture he might appear rather formidable if not positively forbidding, but his utterances were softened by humour. The jokes of fifty years ago, or even of ten, will seldom bear repetition, and removed from their context and robbed of his rolling delivery Lanchester's verbal jests might now seem unduly caustic. His humour and its literary allusions must often have been above the heads of his audience, as when he gave a trial run in the Gold Medal Phaeton to an overbearing and overdressed financial tycoon from whom his directors hoped for an investment. Lanchester chanced to stall his engine in one of Birmingham's crowded streets and thrusting the detachable starting handle into the kid-gloved hands of his passenger he commanded: 'Sartor, re-start us'.

One suspects that many of Lanchester's jokes were private affairs, and it mattered not to him that they were not recognised as jokes at the time. In 1907, for example, the directors had again urged the abandonment of the worm gear and some other distinctive features; Lanchester managed to circumvent these suggestions but agreed to design a more conventional type of car which was to have a T-head side-valve engine, a form of combustion chamber which Lanchester usually condemned in the strongest terms but which appealed to the directors, presumably, because it was currently fashionable. The result was, surely, a jest of gargantuan proportions for the six-cylinder 50hp Lanchester, which made one test run before being hurriedly and deservedly relegated to the scrap heap, violated practically every canon of good design. It was so clumsy and inept (the engine was over 7ft long) by comparison with Lanchester's other work that it must have qualified for the title of the worst car in the world had it ever been put into production.

Lanchester's character, and his fearlessness, are revealed in this passage from a letter he wrote to the consultant engineer to the Director of Military Aeronautics during the First World War:

> ... I want you to be quite clear that I am not writing you from any feeling of resentment that you habitually pooh-pooh my own works as unreadable. You may continue to do this, it amuses me and one or two others, but you really must not ask the committees with which you are associated to spoon-feed you with information which you could obtain if you used the ordinary industry that most of us use in reading over other people's work. ... You must realise what a gross insult it is to the men sitting on a committee of that kind to suggest that *your* time is too valuable to allow you to read what has been written and what is available, but that their time is usefully employed in doleing out the information to you verbally and in abstracts. ... If there were others like you the committee would have to become a kind of scholastic establishment to teach elementary science to students, and not always very apt or profitable students at that. ...

It is small wonder that knighthoods and OBEs did not come his way: every generation needs a Dr Lanchester but ours, alas, is without one.

3 HENRY M. LELAND

by MAURICE D. HENDRY

HENRY LELAND (1843–1932). American: Precision engineer and machine-tool designer who became involved with motorcars in late middle age. With Cadillac 1902–17, famous for the Cadillac V-8 of 1914 and the first Lincoln V-8 of 1920.

HENRY M. LELAND

HENRY LELAND MADE micrometers and machine tools, automobiles and aero-engines, but above all he built *men*. Even the giants of the American automobile industry learned from him and held him in awe. Fred M. Zeder, chief engineer of Chrysler, said: 'We called him the Grand Old Man of Detroit. He was indefatigable and so patient in his directing and guiding.'

Alfred P. Sloan, creator of the modern General Motors, recalled: 'He was fine, creative, intelligent. Quality was his god. I regarded him as my elder, not merely in years but also in engineering wisdom.' Even that arch-autocrat, Henry Ford, admitted to holding Leland 'in profound respect'.

Leland was almost a generation older than the majority of the engineers of his day, but seniority alone did not account for their deference. As a self-made man, he was assured of admiration for a start, and this impression was heightened if the admirer was fortunate enough to meet him personally, for he was imposing in stature and patriarchal in appearance. His black-or-white moral code commanded further regard—even those who could not tolerate his puritanical beliefs still respected him for them. But above all, it was what he symbolised that caught the imagination. His heritage was the unsurpassed skill of the great New England machine-tool builders, a heritage which he had taken and spread throughout the country. During his lifetime—because of him and others like him—the United States became the most powerful industrial nation in the world. To the young Detroit mechanics and 'barn engineers' at the turn of the century, Henry M. Leland was The American Machine.

Henry Martyn Leland was born to Leander and Zilpha Leland in a farmhouse near Barton, Vermont, on 16 February 1843. The Lelands had been in America for nearly two hundred years—the first Henry Leland had landed in Massachusetts in 1652. Leander Leland had inherited land from his father, but had lost it through an ill-starred venture into cattle trade. Henry was the eighth child, the farm was heavily mortgaged, and Vermont is a hard country, so frugality and the will to survive were nurtured in Henry Leland from the beginning. So also were high moral standards, both his parents being sincerely religious. The emphasis was on practical Christianity rather than pettifogging dogma—square dealing, kindness and assistance to others were the keynotes. In work, the children were firmly instructed: 'There is a right and a wrong way to do everything. Hunt for the right way and then go ahead.'

Henry applied this principle early. At the age of eleven he had the opportunity to earn some money for the family by taking on work put out by the shoe factories. Pegging soles, he worked out a method that allowed him to peg fifty pairs a day. At 3 cents a pair, the schoolboy was earning $1.50 a day when the best men in the trade were doing no better than $1.75.

In 1857, Henry's family moved to Worcester, Massachusetts, where he joined the Crompton-Knowles Loom Works as an apprentice at the age of fourteen, working a sixty-hour week for what was then considered good pay—three dollars a week. He had an interesting encounter with a mechanic at Crompton-Knowles, who said to him one day, 'Thou'll never be a mechanic, lad', thereby proving that it never pays to prophesy.

Britain was the hotbed of the Industrial Revolution and the world pace-setter at that time. There were already signs, however, of the awakening American colossus. In the field of arms manufacture, for example, the USA had surpassed Britain. Starting with the pioneer efforts of Eli Whitney in 1798, the New England armourers, Whitney himself, Simeon North, John Hall, Robbins and Lawrence, and Samuel Colt had developed a principle that was to have profound influence on the world—and in no place more so than the yet unborn motor industry. The principle was interchangeable manufacture. Its practitioner in the automobile world would be Henry Leland. The hyphenating factor between the man and the idea was war—the American Civil War.

Leland was still at Cromptons finishing his apprenticeship when the war started. He was a first-class rifle shot and a supporter of Abraham Lincoln, so he attempted to join the Union Army, but was rejected as under age.

At Cromptons, a war-priority job had to be done. The Springfield Armory asked the factory to supply a Blanchard lathe, invented by Thomas Blanchard from Massachusetts. It was a precision tool for copying irregular shapes suddenly in demand—gunstocks. Most of the factory's men had left, and the delicate job of building the lathe would have to be done by the best apprentice. The superintendent selected 'the lad who would never make a mechanic', and the job was done to perfection.

Leland went with the lathe to the Springfield Armory, where he could learn more about the new system of interchangeable manufacture. He worked there until the end of the war in April 1865, when, along with many others, he was laid off. The next day he joined another armory whose name was already internationally known—the Colt Revolver Works in Hartford, Connecticut. This factory was probably the most highly organised and lavishly equipped machine shop in the world at that time, and was under the direction of Colt's superintendent, Elisha King Root, then considered the finest mechanic in New England. He had personally designed many of the 1,400 machine tools in the factory, planned their installation and trained the staff. The Colt Armory had already begun to influence American industry; some of its 'graduates' were A. F. Cushman, founder of the Cushman Chuck Company, and that world-famous pair, Francis Pratt and Amos Whitney.

In the Colt Armory, Leland learned that machine tools could be designed to supersede even the smallest manual work and do it better. Even the burrs on machined parts were removed by Root's special machinery, and small forgings were die-stamped perfectly accurately in large quantities. The comprehensive array of jigs, fixtures and gauges ensured the accurate manufacture of all parts to a rigid standard and made them completely interchangeable. The initial cost of all this equipment was very high and had never been attempted on such a scale before. Other arms makers who closely followed the Colt project predicted failure; instead, the plant doubled its size within six years.

All this registered in Leland's mental file. He already had a passion for precision and had gained a reputation as an outstanding workman during his two years with Colt. When he married and moved to his wife's town, Worcester, Massachusetts, he worked at several other tool and arms works. Then, in July 1872, he took another decisive step in his career, when he joined Brown & Sharpe at Providence, Rhode Island. The factory was already long established and had produced the first practical hand micrometer and the first universal milling machine. These and their other products were in demand all over the world because of their advanced design and the utmost precision of their manufacture. Here, men worked to incredible tolerances—as close as four *millionths* of an inch. Leland was at once at home in this environment, and became a toolmaker in the gear and cutter section.

Many improvements at Brown & Sharpe arose from his now budding genius. An order for horse clippers resulted in Leland making a close study of the product. He then evolved an improved version, lighter and more efficient, suitable for human hair. Soon the firm were making 300 a day. He then put forward suggestions to Mr Joseph Brown, head of the firm, for a design of grinding machine. Brown, himself, had had such a machine in mind in 1868, but had put it aside. After several years, and incorporating ideas from Brown, Leland and other members of the staff, the finished article appeared in 1876 as the world's first universal grinder. It was an immense step forward, the parent of all precision grinding machines, and recognisably the same tool today. (See drawing, p 86.)

Precision grinding now became widespread. Eventually the original Brown & Sharpe grinder proved too light for its work, and in 1886 Leland and Charles H. Norton redesigned the grinder, making it heavier, stepping up the speed and power, and increasing the coolant supply.

The Browne & Sharpe Universal Grinder, dating from 1876, was the foundation-stone for the precision grinding machines that made possible interchangeable parts in automobiles. Leland was not only its most enthusiastic advocate, he also took part in its design, development and promotion

In 1878, Leland had become head of the sewing-machine department, succeeding Richmond Viall, who had been appointed general superintendent of the whole Brown & Sharpe plant. Leland gave an undertaking to Messrs Sharpe and Viall that, if allowed a free hand, he would guarantee substantially to improve the efficiency of the department within one year. This was agreed upon, and when the end of the year came and careful audit was made, it was found that labour costs had been reduced by 47 per cent. Leland later summarised the thought and reorganisation which had gone into achieving this result in a paper entitled 'The Art of Manufacturing'. In this paper there were such key phrases as 'it is the foreman's place to know that every piece of work turned out by his department is RIGHT, and it is his work to teach his men how to make them RIGHT. . . . it doesn't cost as much to have the work done RIGHT the first time, as it does to have it done poorly and then hire a number of men to make it right afterwards'.

Leland was now forty-one, and an accomplished engineer, designer, production man, efficiency expert, technical adviser and, thanks to his interest in politics and his fellow workmen, social philosopher as well. Sharpe and Viall asked him to represent the firm in the burgeoning area of the mid-west. Recognising him also as a unique asset, they gave him a liberal hand. He was not required to bring back orders, and he advised customers without restriction from his own management. One rival salesman was shocked to hear Leland once advise that 'Pratt & Whitney build the best machine for this job'. But Brown & Sharpe realised that Leland's impartial advice bespoke a complete integrity that paid off for them in the long run. The manufacturer who ordered the Pratt & Whitney tool also placed a heavy order with Brown & Sharpe, realising that he could not do better than patronise a firm who employed men of this calibre.

One of the most important manufacturers to ask his counsel was the Westinghouse Company. Their air brakes had revolutionised railroad practice, but the need was for greater precision in manufacture. A good deal of hand scraping and adjustment was needed to ensure efficient operation of the pistons and cylinders, and the results were still sometimes temperamental. Leland demonstrated the superiority of the Brown & Sharpe grinding process by returning to Providence with a set of pistons and cylinders which he personally finish-ground on improvised equipment. Westinghouse placed a substantial order.

In 1890, Leland at last realised his dream of going into business for himself. He had saved some money from his now substantial salary, and found financial backing in Detroit through Charles Strelinger, who sold machine tools and hardware and knew Leland well. The backer was Robert C. Faulconer, a wealthy lumber man. Detroit was an obvious selection, strategic and expanding. Charles Norton, Leland's associate at Brown & Sharpe, joined him and Leland, Faulconer & Norton, with a total capital of $50,000, was organised in Detroit in September 1890. Faulconer supplied $40,000, Leland $1,600 plus a further $2,000 loaned to him by Brown & Sharpe. Faulconer was president, Leland, vice-president and general manager at $2,000 a year, Norton—later a famous manufacturer of crankshaft grinders—was machine-tool designer, and

Strelinger was secretary. The main work was gear grinding and special tool making. The firm did well from the start, and within three months its employees rose from the initial twelve to sixty.

Leland's son, Wilfred, had been working part-time at Brown & Sharpe to help pay for his medical studies, and proved to have the same superlative mechanical skill as his father. He decided to specialise in engineering instead of medicine, and covered in one year virtually the entire apprentice's course normally occupying three years. At the end of this one-year pressure course, as a demonstration of his prowess, he made with his own hands a set of master gauges accurate to one-hundred-thousandth of an inch. Wilfred joined his father's company and was put in charge of the gear-cutting section. With two Lelands now in the plant, the guiding genius was referred to by his colleagues and workmen as 'HM'.

The firm produced tool and lathe grinders, milling machines and gear cutters of their own design, and acted as engineering consultants to inventors and local industry. After three years they built a bigger factory, and by 1896 the capital stood at $100,000. In the same year they built a foundry for high-grade casting work. This was so swamped with orders that the harassed superintendent relaxed inspection standards to try and raise the output of castings. Immediately Henry Leland intervened, and for months personally inspected every batch. Casting after casting was thrown out until the rejection rate approached 50 per cent, even though many of the castings would have been passed by the average foundry. The whisper went around 'The old man is going crazy', and complaints reached Faulconer from all directions. He attempted to persuade his general manager to compromise, but Leland remained adamant, retorting 'There always will be conflict between Good and Good enough . . . one can count on meeting this resistance to a high standard of workmanship. It is easy to get co-operation for mediocre work, but one must sweat blood for a superior product.' He had his way, and the company continued to receive more orders than it could handle even at prices far above its contemporaries. Regular customers found that although a Leland casting might cost twenty cents a pound against an average of eight cents, the quality was superior, and less machining was required because of the close tolerances held.

During the American bicycle boom of the 1890s, gear-driven bicycles were popular. The leading makers, Pope of Connecticut, and Pierce of New York—both later to become automobile manufacturers—had trouble producing satisfactory gears and approached Leland. He showed them that their hardening process was faulty and produced inaccurate

Page 90 (above) The 1908 Cadillac 'Thirty', which the Lelands advertised as 'the most accurately-manufactured car in the world' with more than 400 operations held within limits of one-thousandth of an inch; (centre) a 1919 Cadillac V-8 town car. Stripped of body, it won the first New Zealand Motor Cup at an average speed of 88 mph in 1921;

1918 Cadillac V-8 chassis; note the Delco distributor and : in the engine vee, detachable cylinder heads, unit engine-gearbox and full dual exhausts

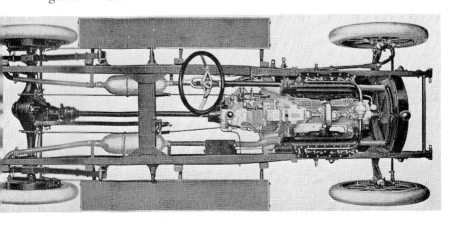

Page 89 (*above*) Henry Leland with his protégé Kettering and
associated with the development of Cadillac's early electrical equipr
J. B. Edwards, president of Kellogg Switchboard & Supply Co;
Delco engineer; Henry M. Leland and Charles Kettering; (*belo*
Cadillac: a Leland engineering conference in 1902. Left to right:
Walter Phipps, Ernest Sweet, Wilfred Leland (back to camera) and

(*below*)
dynamo

wheels and teeth profiles. For them he developed a hardening and profile-grinding process and a special gear generator that enabled the production of vast quantities of gear sets, accurate to a half-thousandth of an inch and interchangeable from one bicycle to any other. Pope and Pierce placed orders for the Leland gear cutters, and Leland's firm began making thousands of gear sets for other bicycle manufacturers. In 1896, the company began making steam engines, hundreds of which were used on Detroit tramcars. The same year Leland took up the manufacture of the internal-combustion engine for marine use in a range from 5 to 20hp. These were the first petrol engines in the world manufactured on the principle of interchangeable parts.

Leland's involvement with the automobile industry proceeded logically from this point. He was eventually to found two separate makes —but neither was to carry his name. He was led into automobiles by another man who also created two makes—and who got his name on one and his initials on the other.

Ransom E. Olds came to Detroit in 1899 and set up a large—for the time—factory to build his famous 'curved dash' model in quantity. He had already built several steam- and petrol-powered vehicles in Lansing, but not in serious production. Production of the 'curved dash' Olds began in 1900. One weakness was its noisy transmission, and Olds called on Leland's advice. Leland was appalled to find that the teeth of the gears were hand-filed by mechanics to make them fit. He soon supplied Olds with a quiet-running transmission in which all gears were precision-ground and interchangeable car to car without any hand fitting. Shortly afterwards, Leland & Faulconer were given a contract for 2,000 engines for Olds, and along with the Dodge Brothers, they became the principal suppliers of Olds engines. The design was identical but there was a subtle difference—which gave a not-so-subtle superiority in performance. The first to make this public seems to have been a man named Henry Ford. Attending the first Detroit Automobile Show in 1901, he saw an Olds exhibit designed to show consistency of engine performance. Two engines were running alongside one another with identical throttle and valve settings. Two dials indicated the same speed. Ford, perceptive always, suspiciously examined the engines closely. One was running with a brake load on its flywheel. Ford pointed this out to the Lelands when they attended the show and Wilfred Leland later recalled:

'My father and I were somewhat amused to find that the braked motor had been manufactured by us, while the other motor had been built by the Dodge Brothers. The Dodge-made motor delivered 3hp; the Leland-made motor 3·7, due entirely to closer machining.'

F

For Ford, the demonstration had a profound meaning. At the time he was personally unknown to the Lelands, but he never forgot the incident. Two years later, when Ford was making his third attempt at automobile manufacture, he asked Leland's advice on grinding pistons. Leland was not, however, impressed by Olds's methods. Olds took a rough and ready view of research and testing in general. His attitude was that it was time enough to correct a fault when it occurred on the road. To Leland, such a philosophy was anathema. He set about improving the Olds engine, and Wilfred recalled:

> 'We made larger valves, held the valves open longer, instituted the efficient timing we had learned on our marine motors, and the motor we developed delivered 10·25hp. We offered this motor to the Olds Company, taking it down personally to Fred L. Smith, the business manager. We emphasised that this motor, the *same size* as the other, would develop nearly three times the horsepower. We could manufacture it at *the same cost*. We were dismayed by Smith's refusal to use it. He was unwilling to bear the cost of retooling for new parts in the rest of the car.'

But another firm was soon keen to use Leland's engine. This firm had been founded in 1899 by William H. Murphy, with Henry Ford as his engineer. It had started as the Detroit Automobile Company, but surrendered its charter in November 1901. It was revived a year later as the Henry Ford Company, but Ford had mainly concerned himself with building two racing cars and had made little progress toward actual production other than designing and building a single prototype. After disagreements, Ford left for good in March 1902. In August, the directors approached Leland for an appraisal of their plant machinery for liquidation. Instead, he appeared dramatically at the directors' meeting with his new engine under his arm, and urged them to stay in business. Besides the engine, he could offer them his forty years' experience in manufacture. The directors asked if he would reorganise their firm. The offer was accepted, and on the way home Leland twinkled, 'I'll be sixty next birthday, but I believe I could get a job as a salesman yet.'

Capitalised at $300,000 the new Cadillac Automobile Company (named after the French explorer who had established a trading post at Detroit 200 years earlier) was established in August 1902. Leland became a stockholder and director as well as being technical adviser, and Leland & Faulconer were awarded the contract for engine, transmission and steering gear for the new car. At first the Leland components were delivered to the Cadillac factory for assembly into chassis and body supposedly waiting. Frequently, however, the Leland & Faulconer units did the waiting—the chassis plant was unable to keep up with the methodical regularity of engine delivery. So Leland took over the whole operation.

The first Cadillac was designed by Ernest Sweet, Frank Johnson and Alanson P. Brush under the supervision of Henry and Wilfred Leland. An engine-testing room was laid out and equipped by Charles Martens who, to his death in 1960, took great pride in having been selected personally by Henry Leland to supervise every dynamometer room in all the Leland-managed plants. All these men had been trained by 'HM' personally and had already worked for years at Leland & Faulconer. The Leland philosophy of design (despite the genius of 'HM' himself) had no place for the 'all-purpose genius' romanticised by journalists but more often fiction than fact. Instead, the method was in principle similar to modern 'team' practice. Leland explained:

> 'Cadillac cars are not the exclusive design of any one person. They represent the composite ideas of a number of inventors, designers and engineers, each skilled by many years in his special branch of work. Every feature of Cadillac cars is thoroughly considered by a special committee of mechanical experts. No feature is adopted until . . . fully proven by long and severe tests.'

The basic design of the car was obsolescent by European standards, but perfectly logical for the market in view—reliable and comfortable travel in 'frontier' conditions. These were, after all, characteristic of almost all the territory centred on Detroit, as well as Canada, Mexico, and many overseas markets as well. For these areas, the Americans had evolved the 'buggy-runabout' concept, with flexible suspension, large wheels and high frame with plenty of ground clearance, overall light weight, and a simple, reliable single or twin engine with planetary two-speed transmission and chain drive, capable of absorbing repeated abuse. Olds, Ford, and others had built cars on these lines, but they were buggies with substantial bugs. Their designer-builders in several cases were men of great potential as they later proved, but initially they simply lacked Leland's unmatched experience in machinery. The first Leland-built car, on the other hand, although differing little in design, was markedly superior in execution. He had improved almost every detail, ranging from the use of exceptionally high-quality alloy steels in frame, axles and other components, to such striking details as the specially-designed self-locking nuts used throughout. The car was an immediate success, and for a period of years Cadillac was the largest-selling make in the world. Undoubtedly the greatest event in the career of the single-cylinder model, which was continued until 1908, was the winning of the Dewar Trophy, as a result of the famous Standardisation Test at

Brooklands. This has been recounted so often that there is no need to go into the details here. What was important about the test can be summarised as follows:

(a) It was designed specifically to prove or disprove the possibility of building cars with interchangeable parts.

(b) The conditions were laid down by an impartial and expert tribunal composed of engineers of the Royal Automobile Club of Great Britain.

(c) The test was held in a country that was not merely neutral, but actually critical of American methods of manufacture—Great Britain.

(d) The cars were completely standard, and were selected at random by the RAC.

(e) The test was open to all automobile firms but, despite claims by other makers that they built cars with interchangeable parts, no one but Cadillac was prepared to submit to the test.

Leland's reaction to the Dewar Trophy award was typical. The cup itself, a magnificent silver trophy of great size and value, was placed on display in the factory where all the workmen could examine it. Each employee was given a small leaflet describing the test, with 'HM's' personal congratulations—'the honour belongs equally to every honest, sincere and conscientious member of this organisation, no matter what his position, who has striven constantly and patiently to acquire and maintain in the work he is doing each day that fine accuracy which has made possible the absolute interchangeability of parts in Cadillac cars.'

As each component supplier became associated with Cadillac, so the gospel of interchangeability was spread. This is confirmed in the memoirs of representatives of the Timken and Hyatt roller-bearing concerns, one of whom was put on the carpet by Leland for supplying bearings showing wide variations in sizing tolerances. The other witnessed the 'carpeting' of a manufacturer of humble grease cups which failed to pass thread-gauge tests at Cadillac. Leland's crusade was greatly assisted by the work of his former associate, Charles H. Norton, who was now producing heavy precision-grinding machines of very advanced design, exactly suited to the automobile industry.

Since the industry's growth was outstripping the supply of skilled machinists and toolmakers, in May 1907 Leland opened the Cadillac School of Applied Mechanics, 'for the benefit of the community, the industry and the country as a whole.' Here trainees were educated in all phases of machine work, automobile design and assembly, drafting, mathematics and metallurgy. This school, the first of its kind in the US automobile industry, was later emulated by Henry Ford with the Ford Trade School, Ford being publicised as the originator of the idea.

Despite the great success of the single-cylinder car, the Cadillac engineering group were well aware that this vehicle would be but a passing phase, and that as Detroit became urbanised and roads improved everywhere, the next step would be a European-type car. Accordingly, as early as 1904, only a year after they had put the 'single' into production, 'HM' and his engineers had begun on a four-cylinder car, with three-speed and planetary transmission shaft drive. This appeared in 1905 as the Model 'D', and continued until 1908 in 20 and 30hp guise.

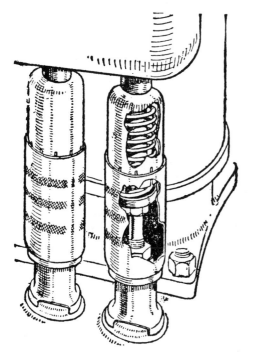

At a time when even the most expensive cars exposed their valve gear, Leland's attention to detail is shown by this neat enclosure by telescopic tubes on the four-cylinder Cadillac

Fine but expensive, these models did not generate enough sales in relation to plant investment to be the mainstay of the company when the single was phased out; consequently, in 1908, Leland's team had an all-new four ready which replaced all previous models. This car illustrates Leland's inherent caution and respect for soundly based, step-by-step progress. Many reputations had been made and lost in the industry because makers failed to gauge the market for a new model. The Cadillac Thirty was carefully planned to meet the needs of a clientèle whose requirements had outgrown the single cylinder. It was a lot more car but it did not cost proportionately a lot more to buy, because the tooling, the profit margin and the market had been carefully worked out in advance. The car presaged a dramatic step which Cadillac would take a few years later—the permanent aim to combine the highest quality with a substantial volume of business. Leland knew that the superior standard of living of the American workman depended on far higher wages than ruled in Europe, where skilled men were employed at virtual slave-labour rates. He also saw that even in the high-price field, bespoke methods of car manufacture would not endure. He therefore rejected the idea of ultra-limited production with fancy prices and produced a quality car which, though it sold for only $1,400, yet had standards of manufacture equal to or surpassing cars of several times its price.

Efficient production was the key to this exceptional value. Not only was Leland's plant laid out to take advantage of the most advanced machine tools, but it was consistently kept up to standard, even though this meant steady scrapping of tools and expensive re-investment in improved machinery. This made possible quantity production while maintaining first-class workmanship, and a high unit rate of profit was unnecessary because of the relatively large quantity of cars produced—about 8,000 annually. Cadillac continued as one of the most prosperous companies in the industry.

While, in engineering matters, Leland was like a mechanical Socrates, in local politics he was like a fiery Old Testament prophet. He took intense interest in social welfare and civic matters, and time after time would arrive at the office fiery-eyed, clutching a newspaper and ready to call out the office staff to 'save the country'. He organised the Detroit Citizens' League, served as its first president for many years, and did much to improve local government and correct abuses; so much so that a contemporary judge once publicly labelled him, 'Detroit's best citizen'.

By now, 'HM's' son, Wilfred, was second only to his father in Cadillac affairs. He had inherited his father's mechanical ability and his reverence for order and method, but in other respects the two men were com-plementary rather than comparable. Physically they were dissimilar, Wilfred being shorter and nowhere near the powerful build of his parent. This difference was reflected in their mental make-ups. The elder Leland was outspoken, decisive and quick-tempered, inclined to see things as black or white. He loved mechanical statistics but detested finance and accounting. On a rare occasion he had been known physically to shake sense into a man who was being particularly difficult. Wilfred, on the other hand, was diplomatic, smooth-spoken and handled people gently. From the age of fifteen he had capably done all the family budgeting, and he showed outstanding talent as treasurer both at Leland & Faulconer and at Cadillac. He drew up a schedule of supplies, production and potential sales for launching the Thirty that proved remarkably accurate, and was shortly to show even greater mettle. Father and son always got on well together. 'They loved working with one another,' recalls Mrs

Ottilie Leland, Wilfred's widow. Closest to the two Lelands was 'HM's' consultant, Ernest E. Sweet, the trinity being referred to in the shops—though not to their faces—as 'Father, Son, and Holy Ghost'.

William C. Durant, busy putting together the original General Motors (later re-formed by Sloan), had selected Cadillac as the top car of his line. When the famous promoter decided he wanted something, the realisation was usually a mere formality. But bagging this quarry proved to be another matter. Beneath his mild exterior, the quiet, gentle Wilfred proved to be one of the most redoubtable negotiators Durant had ever faced. The bargaining and manoeuvring went on for nearly a year, during which Wilfred used the prestige of the Dewar Trophy award to boost the value of the company. Eventually, after Wilfred had forced Durant up one million dollars to a final valuation of $4,500,000, the Lelands entered General Motors on their own terms—a free hand to continue running Cadillac exactly their own way. 'You will receive no directions from anyone,' Durant had told them.

The principle of General Motors was the same then as now—a vigorous combine with control of its own suppliers and a car for every purse. But because of Durant's erratic methods, the organisation soon found itself in serious financial trouble. It was fortunate for Durant that he had brought Cadillac into the family. Buick, with its volume sales, had been the base for establishing GM in the first place, but in the financial crisis of 1909–10 it was the stability of Cadillac that carried the whole group through. Buick had a debt of eight million dollars, Cadillac was solvent and debt-free. At the time, 'HM' was visiting automobile plants in Europe, and again Wilfred played a key role. At a special meeting held in New York in September 1910, General Motors' bankers summoned Durant and his executives to account for GM's finances. As each unit of the corporation was investigated and its indebtedness and casual accounting revealed, the bankers became more and more caustic and determined to cut their losses by liquidating GM. Cadillac was taken up last, and for two hours Wilfred presented the Cadillac statement, answered questions and discussed GM in general. Surprised and impressed, the bankers asked him to wait behind when the meeting adjourned at 6 pm. Durant and the other GM executives left in despair, convinced that the morning would bring the dissolution of General Motors. Instead, when they resumed at 10 am next day, they were astonished and relieved to find that Wilfred had, on the showing of Cadillac, won the bankers over.

In their building process, the Lelands imparted to the other GM units their methods and techniques, either personally or through their key associates. 'HM's' foundry superintendent, Joe Wilson, sorted out

Buick's foundry problems, while his factory superintendent, Walter Phipps, advised Oakland and Buick on production. Ernest Sweet, Leland's most capable engineer, made frequent trips through the other units, advising, guiding, and reporting back to Henry. Charles Oostdyke, Wilfred's purchasing agent at Cadillac, was appointed purchasing agent over all GM divisions. Henry Leland himself personally taught the key machinists to use precision gauges, inspection and testing methods. Charles Nash and Walter Chrysler, then at Buick but later to found their own companies, spent weekends at the Leland country home discussing manufacturing problems. 'HM' even visited the Fisher body plant, recommended a change to metal-panelled bodies and other improvements and placed an unheard-of order for 150 units to start off quantity production of closed bodies—a new development in this field.

During all this frenzied activity, 'HM' showed amazing endurance for a seventy-year-old. His office was open day and night. Motor assembly mechanics would be astonished to see him in their department at two or three in the morning. Later still, the night-watchman would find him asleep at the drawing-board. Yet in the morning he would be as fresh as ever.

Technically, Cadillac repeated their Dewar Trophy triumph in 1914, this time in the area of design instead of manufacture. In most areas, Europe had led in design while America had led in manufacture, but the gap was narrowing and, in fact, Cadillac was shortly to take the lead in both respects. In electrics, they were already ahead, due in no small measure to the encouragement Leland had given someone he termed 'a young and generally unknown electrical genius'—Charles F. Kettering.

Kettering, an engineering graduate of Ohio University, was in charge of the Gilbertian-sounding Inventions Department at National Cash Register. Since 1908, he had been working on coil ignition—an old principle that had been displaced by the high-tension magneto. By a combination of innovation and development, Kettering transformed coil ignition into a system more than competitive with the magneto. Key points of his 'Delco' ignition were the ignition relay and the rigid, spring-loaded breaker arm in the distributor. The former was completely novel, the latter an improvement amounting to a new principle. Previously, a rapidly vibrating flexible blade—the wipe contact—had been used and gave a continuous but weak shower of sparks. Kettering's system produced a single, intense, accurately-timed spark. To prevent arcing at the points, he used an efficient condenser. The overall result was a vastly better coil ignition than had been known before.

Authorities differ as to the installation of Kettering's system on the Cadillac. Kettering's biographer, and others, stated it went into the 1910 models. But Mrs Wilfred Leland maintains that the 1910–11 Cadillac ignition was designed by Leland's own electrical specialists headed by Clair Owen, a graduate electrical engineer.

There is general agreement that Kettering and Leland came into contact through a mutual association—Earl Howard. He had worked with Kettering at National Cash Register before becoming assistant sales manager at Cadillac. But while the Kettering version has Howard linking the two giants via Leland's interest in the new Delco ignition, the Leland account states that it was the self-starter that brought them together and, incidentally, challenges the long-standing belief that Kettering invented the self-starter. The famous incident occurred as follows.

Leland's friend, Byron Carter of the Carter Car Company, had died after injuries suffered in 1910 while cranking a stalled car. Ernest Sweet had taken Carter to hospital, and the fact that the car was a Cadillac added to Leland's distress.

'Those vicious cranks!' he burst out. 'I won't have Cadillac hurting people that way.'

That morning, the usual engineering conference presided over by Leland had, in addition to Frank Johnson and Ernest Sweet, Cadillac's electrical expert, Herman Schwartz, and Fred Hawes, a talented engineer on whom 'HM' greatly relied. They were instructed by Leland to put other work aside and to begin at once on a self-starter. After examining the few starters then on the market, and as a result of observing the home generating set on Wilfred's estate, where a four-cylinder Cadillac engine drove a generator and kept a roomful of batteries charged, they concluded that an electrical motor-generator was the answer. Frank Johnson designed the mechanical components, and when the layout had been worked out, decided that the starter motor was too big. At this point Kettering was asked by Wilfred Leland to produce a smaller and more efficient design out of his experience at National Cash Register Company. The key principle here was that the motor could be safely overloaded, since it was only used for short periods and had plenty of time to cool off. This was the same offbeat approach Kettering had used in designing the National Cash Register machines. So the Cadillac/Delco self-starter was born, and first demonstrated on 27 February 1911. The next step was to include electric lighting, and here Kettering supplied the key feature with his mercury voltage-regulator, which solved the charging problem. (See drawing, p 100.) With the help of Cadillac engineers, the complete system went into production on the 1912 Cadillac. Kettering found Leland to be 'a most exacting and critical purchaser' who 'found fault with many things'. But once the bugs were eliminated, Leland had complete confidence in him, in fact Kettering

had been astounded at Leland's decision to make the Delco system
standard on the entire Cadillac line. Later, Leland told him, 'I really felt
sorry for you fellows, but every time you did something more, the
system got better.'

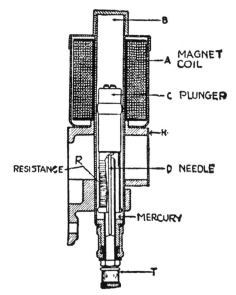

The Delco mercury-type
voltage regulator was specially
designed by Kettering to
Leland's requirements in 1911.
Plunger C was raised or
lowered by magnet coil A,
varying the resistance of the
field circuit and increasing or
reducing the current according
to the state of the battery
charge

There have been attempts to debunk Cadillac claims in electrical ad-
vancements on the score that Cadillac was not the first to use coil ignition, self-starter, or even electric lighting *individually*. This is mere
technical hairsplitting. What is important, and what critics fail to com-
prehend, is that Cadillac was the first make to have a completely successful, self-contained electrical system performing the three functions—
starting, ignition and lighting. Furthermore, this was standard equipment. No other manufacturer had done this before. From this point
onwards, other starting devices became obsolete, the magneto declined,
and acetylene lights faded.

But despite its leadership in electrics, highlighted by the unique honour
of a second Dewar Trophy, the four-cylinder Cadillac was now obsolescent. Six-cylinder makes in the USA had increased to near parity
with fours, which had declined by 20 per cent during 1911–13. Pierce-
Arrow, Peerless, Stevens-Duryea and Franklin had all built very successful sixes. But many others were badly designed and had trouble with
spindly crankshafts, periodic vibrations, and poor carburation. 'HM' and
his engineers designed a six-cylinder replacement for the four, taking
care to eliminate these shortcomings but, even so, Wilfred commented
'I do not like the six crankshaft in principle. It is too long and heavy.'

He argued with 'HM', 'If six cylinders give the same power with lighter impulses than the four, then eight still smaller cylinders will give still lighter impulses than the six, and because of the lighter moving parts, the eight can be run at higher speeds than either sixes or fours.'

Wilfred's father soon accepted the idea of the V-8, and the experimental six was discarded. A number of V-8s had been built in both the USA and Europe over the previous ten years, but the only one to go into production was the De Dion in 1910. The first series De Dion V-8 was a poor design and unsuccessful, but in 1912 the French firm brought out a much improved engine. The Cadillac design was under way before this De Dion appeared in the USA; nevertheless, the second French engine influenced the Cadillac design staff. The trouble with the De Dion was that it was half-baked, a brilliant design with serious shortcomings even in its improved form. Nor did the De Dion organisation realise the potential of what they had, grasp the initiative, and develop and promote the car. They were already a dying concern, not to be compared with the dynamic Leland organisation.

Elaborate care was taken to keep the V-8 project a secret. Supervised by the Lelands, Ernest Sweet and Frank Johnson retained their key engineering positions, and the design staff had as chief draughtsman D. McCall White, a brilliant man who had formerly been with Napier in England. Careful analysis of the 26hp De Dion showed that its basic structure was sound, but that the performance was disappointing. Restricted induction passages and small valves limited the power output. Furthermore, the cooling system would not have permitted sustained high power anyway. The exhaust layout, too, was bad. It also had magneto ignition, which was immediately rejected. And so on. Endless testing, improving, re-designing, re-testing and perfecting continued for two years.

The Leland engine that finally emerged in September 1914 still bore a superficial resemblance to the De Dion, but actually was vastly superior. Just as Leland had transformed the Olds's engine fourteen years earlier, so his team had now taken another sedate unit and, using the same painstaking methods, achieved a new level of efficiency. The largest De Dion V-8 of the period (6 litres displacement) developed 50hp at 1,500rpm. The Cadillac V-8 (5·14 litres) put out 70hp at 2,400rpm. Yet it was a simple L-head design. The high power came from the relatively high compression made possible by the easier cooling of the smaller pistons in an eight; large easy-flowing induction passages with large tulip valves instead of the obsolete mushroom type used by De Dion; ignition that gave easy starting yet remained efficient at high speed; thorough research into cam design and valve timing; careful routeing of exhaust headers to eliminate waste heat and maintain low inlet temperatures; and a cooling system that was a model of efficiency.

The advanced engine was matched to a unit-constructed transmission with multi-plate clutch and a silent spiral-bevel gear axle. Its standard equipment was lavish—electric starting and lighting went without saying, but there was a host of other features. Engine-driven tyre-pump, tilt-beam headlamps, thermostatically-controlled cooling system, condenser tank for desert conditions give the general idea. (See drawing, below.) Yet it was the price more than the specifications that shook other manufacturers, ranging as it did from $2,700 to no more than $3,350 when it might have been expected to cost at least a couple of thousand dollars more. When the Lelands revealed that their first year's production schedule for the V-8 was not thirteen hundred but thirteen *thousand* units, it was clear that the American luxury-car market would never be the same again. Leland revealed the size of the task in some drawing-office statistics: 1,922 drawings of the car and nearly 11,000 tool drawings, taking 381 miles of paper and 64 man/years of work.

Thermostatic regulation of the cooling system was pioneered by Leland in the 1914 V-8 Cadillac, as shown here. Modern systems are identical in principle. Note the bleed via the cylinder-head and carburettor water-jacket

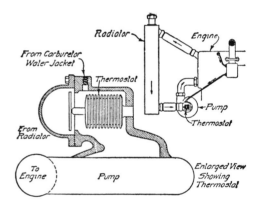

The Cadillac V-8 sounded the death knell of the refined but ponderous and inefficient big sixes which, up to then, had dominated the American luxury-car market. Its flexibility, performance and refinement still impressed W. O. Bentley as remarkable when he recalled it forty years later. Rolls-Royce tested its rear axle to find out why it was quieter than their own. It remained Cadillac's only car for more than a decade, during which production was maintained at the level originally scheduled, around 15,000 annually.

It was a masterly concept. The economics of the car were actually more important even than the technical aspects. How to reconcile high American labour costs with a reasonably-priced luxury car so as to ensure an enduring business, was a problem that would face all makers

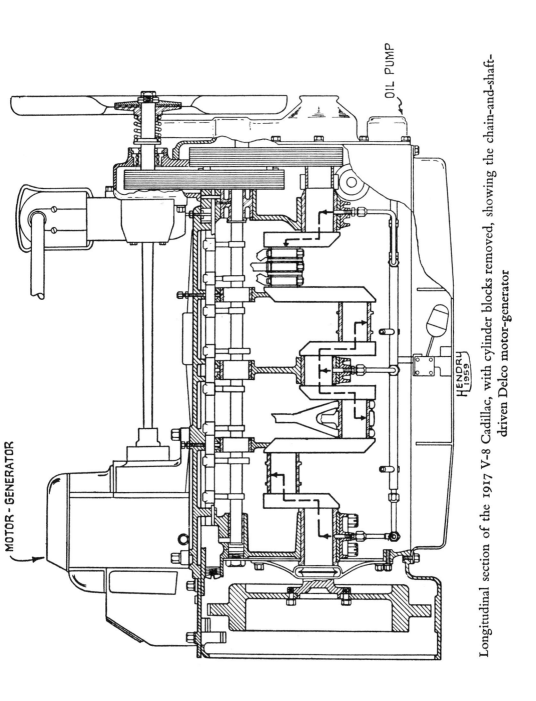

MOTOR - GENERATOR

OIL PUMP

HENDRY
1959

Longitudinal section of the 1917 V-8 Cadillac, with cylinder blocks removed, showing the chain-and-shaft-
driven Delco motor-generator

sooner or later. The Lelands had found the answer. Packard was quick to grasp the meaning of what they had done and shortly adopted a similar philosophy. Pierce-Arrow and Locomobile, however, doing comfortably at the time with very limited production, failed to follow suit and faced serious difficulties in later years.

Establishment of Cadillac in this commanding position, although the last of Leland's major achievements with the firm, still was not the end of his accomplishments. By now he was a world-famous figure, and overseas mail sometimes arrived addressed simply 'Henry Leland, USA'. In 1913, he was elected president of the Society of Automotive Engineers for 1914, and with other engineers he made a second trip to Europe. He returned home worried. Convinced that a war was inevitable, he attempted to arouse his countrymen to the danger, but was ridiculed.

After the war started, Leland travelled to England to offer advice and assistance in production, and to study British aero-engine manufacture. Returning a strong supporter of the Allies, he warned President Wilson to prepare for war, but his advice was curtly rejected. Later, the Cadillac plant received a Rolls-Royce engine from the British Air Board, who asked the Lelands if they could undertake production.

The Lelands, however, preferred to build the new Liberty designed by Col Vincent of Packard. They had supplied two of the engineers and two of the draughtsmen on Vincent's design team, the engineers being Cadillac's electrics specialist, H. G. Schwartz, and Benjamin F. Anibal, later chief engineer of Cadillac. The Lelands approached Durant with the proposition that Cadillac's large new body plant be turned over to aero-engine production. They found him in the same mood as President Wilson. Wilfred had argued quietly but 'HM', in a storm of rage, decided that the only way to get anything done was to resign and start a new concern. Finally Wilfred agreed, and with very mixed feelings they left Cadillac.

In August 1917, they founded the Lincoln Motor Company to produce Liberty engines. The United States was now in the war, and Leland, a lifelong admirer of Abraham Lincoln, drew a parallel with his youth when, more than half a century earlier, he had worked his heart out making guns for the President's army. So the new organisation was named 'Lincoln' instead of 'Leland'.

At this stage the new company possessed nothing material at all, other than a vacant factory site and the ability of its personnel, a number of whom were ex-Cadillac staff who had accompanied the Lelands. Their reputation was solely responsible for the amazing confidence shown by investors, bankers and the US Government, who together furnished twelve million dollars. A contract was signed for 6,000 engines on 31

August, when the Lincoln Company was exactly two days old and no more than a piece of paper. Within six months, the new Leland organisation had built a new factory of 616,000sq ft and equipped it with 6,522 different tool designs aggregating 91,807 in number, and had the Liberty in production. Within ten months, with 6,000 employees, they were building fifty engines a day, equal to the *weekly* production of Rolls-Royce, who had had three years' aero-engine experience and employed 10,000 workpeople. Two million dollars had been spent on tooling—twice that of the nearest competitive Liberty producer—and the result was that Leland-built Liberties were made at a lower standard cost than those of any rival.

The Liberty was a 'magnificent challenge' and Leland—now seventy-five years of age—delighted in mastering it. He said, 'When we were making those engines, I used to work sixteen to eighteen hours a day at the plant, but it was no hardship. Work is the best fun I have.'

Because of his age, there were fears that he might break down. A government commission enquiring into the status of Liberty production questioned the factory superintendent, William Ebelhare, about Leland's ability. Ebelhare, incidentally, was not 'a Leland man' but had been loaned to Lincoln by a Boston engineering firm. He replied:

> 'The conferences he has are more frequent than I have ever seen in any institution larger or smaller. These conferences cover the vital points of manufacturing *and* engineering problems in the motor. There is not a single piece of machinery purchased but he knows its use. He signs the requisition for every piece of machinery; every new process put in I take up with him personally, and sometimes my judgement is reversed. I don't know of any other man in so big an institution as this who covers the factory as broadly as Mr Leland.'

The war contracts terminated in January 1919 with 6,500 engines completed, and the Lelands prepared to build a new post-war automobile—the Lincoln. Under the direct supervision of the Lelands, Ernest Sweet, Frank Johnson and Walter Phipps, several different designs of car were laid out, two of which were finally built in pilot form for testing in 1919. Leland made the final selection and announced the car in September 1920. In essence the new Lincoln was a 'super Cadillac'—a luxury V-8 built to similar standards, but faster and more powerful and embodying all the knowledge gained in the past five years. Experience with the Liberty aero-engine resulted in a novel 60 degree angle between cylinder blocks to reduce vibration, while another 'first' was thermo-static radiator shutters.

In a pre-production advertisement, Leland, twinkling, told of the car's effortless performance with Lincoln engineers on a drive. On a clear road, they were bowling along serenely at about forty—so he judged—when he called to the driver, company tester Harry Marchand. 'Step on it and let's see what she can do.' Told that the car was already all out, Leland was 'keenly disappointed'. But, he added, his spirits rose when engineer Fred Hawes called from the front seat that the car was doing nearly eighty and asked 'HM' if he wanted an aeroplane with a Liberty engine.

'Actually,' Leland commented dryly, 'I am constitutionally opposed to speeding, but my interest and curiosity, I suppose, got the better of me. I believe,' he added, 'that motorists will agree that the ideal car should possess, primarily, six important virtues—good appearance, trustworthiness, long life, power, economy and comfort. The order of their importance is largely a matter of individual opinion.'

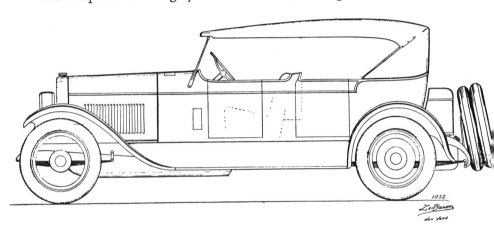

Recognising their own lack of talent for styling, the two Lelands obtained the services of the best men in the field, even before the Ford regime. This well-proportioned, four-seater phaeton was designed for the Lincoln chassis by Le Baron of New York about 1922–3

Fifteen hundred orders had been taken before the public had seen the car, and it had originally been planned to build 6,000 cars the first year, rising to 14,000 thereafter. However, as soon as the car got into production, sales suffered from the onset of the post-war depression and also from the old-fashioned body styles, due to Leland's conservatism in this matter and his preoccupation with the mechanical side. Sales were insufficient for profit and the company required further outside financing until market conditions improved. More stylish bodies were introduced by engaging specialist body builders such as Judkins, Fleetwood and Brunn.

Page 107 (*above*) A group of Leland engineers and executives around the 6,500th Lincoln-built Liberty aero-engine. Left to right: Henry Leland, Charles Martens, William H. Ebelhare, William Guy, Paul Abbot, Ernest Sweet, Wilfred Leland, William T. Nash, Le Roi J. Williams; (*below*) the Lincoln V-8 engine of the 1920s; the last of the Leland designs, it was also the most refined and the most potent

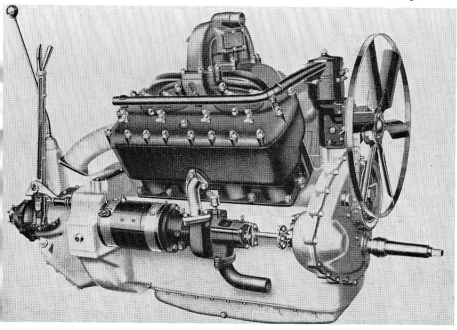

Page 108 (*above*) Creating the Lincoln, a Leland engineering conference in 1920. Left to right: Frank Johnson, Walter Phipps, Wilfred Leland (back to camera), Ernest Sweet and Henry Leland; (*below*) A 1917 Cadillac Roadster. More than half a century old when this photo was taken, it had covered 25,000 miles in rallies and tours, could still top 70 mph and give 12 mpg. It was owned by Mr O.J. Barrett of Johannesburg, South Africa.

As the Lelands were canvassing further finance, a government tax bill for $4,500,000 arrived. This was completely in error, having been wrongly calculated by the Tax Department and was nine times the correct amount. Although the error was later corrected, it ruined the Leland's chances of obtaining immediate financing and, despite their opposition, a board meeting of the company's directors recommended receivership.

Henry Ford bought the company at the receiver's sale for $8,000,000. The Lelands remained with the company, but it became increasingly dominated by Ford, acting through his ruthless chief executive, Charles Sorensen. The Lelands, who were under the impression that they would be running a semi-autonomous company as they had with Cadillac, found otherwise. They left in June 1922, after only five months with Ford.

Misunderstandings and lack of any written agreement are given as major reasons for the break. The Lelands were not beyond criticism, and it is unfair to place the entire blame on Ford. Moreover, it seems unlikely that there would ever have been enough room for both the Henrys in any one organisation. However, the *bulk* of the blame did rest on Ford, as even his most authoritative biographer, Professor Alan Nevins, admits. Ford had serious blind spots—to put it mildly—in his attitude to management. For decades, no schematic organisation existed and Ford ran things like a Turkish sultan, with Sorensen as his Grand Vizier. His 'no-will-but-mine' rule allowed him to dismiss vital men rather than find ways of working with them.

Henry Leland's ghost had the laugh in the end, because ultimately the Ford Company, under the regime of Henry's grandson, was reorganised on General Motors' lines. Even in the 'twenties, Ford found that, although the Lelands were gone, he could not so easily do without their men or their methods, and after some blunders by Ford executives in the Lincoln plant, Thomas J. Litle was made chief engineer. Litle had been in Leland's engineering force both at Cadillac and Lincoln, held many patents, and in 1926 became president of the SAE. When he left Lincoln in 1927 to head Marmon, Ford then persuaded Frank Johnson to come back and take over. Johnson had left with the Lelands, but returned to head Lincoln engineering for the next twenty years. He continued there well past the normal retiring age, until the advent of Henry Ford II in 1947. Under Litle and Johnson, and with the enthusiastic support of Henry Ford's son Edsel, the basic Leland V-8 chassis—with much improved body styling—continued through into the next decade with only evolutionary changes, and Leland ideals were carried forward by Johnson into the fabulous twelve-cylinder KB model of 1932.

G

What manner of automobile were these Leland-inspired Lincolns? Superlative machines. Only a handful of the world's finest cars could compare with them. *Autocar* road testers summarised the V-8 as 'A magnificent town carriage—or an extremely fast touring car with an exceptional performance. A superb vehicle suitable for all occasions.'

Of the V-12, *The Motor* testers wrote:

> A very fine car from every point of view. It is so well thought out in every detail, so well designed, so beautifully made. The highly-finished engine remains perfectly clean and free from oil stains after several laps of Brooklands at 90mph; whatever its speed, the engine (which will turn over at speeds up to 4,500rpm) remains absolutely silent and vibrationless. The acceleration is tremendous, and the Lincoln, despite its size, never gives the driver the feeling he is handling a large and heavy car.

These comments are interesting when considered alongside the fact that Leland never really meant his cars to be mere playthings of the rich, or as connoisseurs' items intended purely for pleasure. He was content to let other makers take the limelight at world salons, become the darlings of the dowagers and battle on the racing circuits. His cars, moulded in his character, were built to serve useful purposes. His Cadillacs crossed deserts and continents to open up the trade routes of the world, while his Lincolns maintained law and order at home. Their attraction to the connoisseur came as a by-product of engineering perfectionism.

The two engineering groups Leland had built up shaped much of Ford and General Motors thinking for years afterwards. Men trained in the Leland techniques were still running Cadillac in World War II. When Ford's successor to the Model T, the Model A, appeared, it looked like a baby Lincoln, and Ford's liking for the V-8 engine undoubtedly grew out of the association with the Lincoln.

After the Lelands left the automobile industry, Wilfred became a mining engineer. 'HM', although well into his eighties, continued active in local affairs for years, attending meetings of the various groups in which he took an interest, and adding to his collection of items on Abraham Lincoln. Always handy with his pencil, he once showed Clarence T. Wilson, a prominent lawyer, sketches he had made for an 'automobile of the future'. This was to have four-wheel, hydraulic displacement drive and braking, a rear engine, four-wheel independent suspension and four-wheel steering.

In 1920, he was awarded an honorary degree as Doctor of Engineering by the University of Michigan, and in 1923 was similarly honoured by the University of Vermont, in his home state. He died on 26 March 1932, aged eighty-nine.

Leland's impact and influence were of a similar order to Ford's, in fact a record of the 'Two Henrys', and of the men and companies associated with them, could constitute an outline of American automobile history. A Ford booklet printed some years ago called Leland 'an iron-willed engineer in pursuit of perfection', but probably the pithiest comment on his ideals and work is the slogan he began at Cadillac and which is still used in that make's advertising:

'Standard of the World'.

4 HANS LEDWINKA

by JERROLD SLONIGER

HANS LEDWINKA (1878–1967). Austrian: Nonconformist designer for Nesselsdorf, Steyr and Tatra in Austria and Czechoslovakia. Advocate of backbone chassis, all-independent swing-axle suspension, air-cooled engines. Produced a V-12 front-engined car for Tatra in 1927–8, and in 1934 his first rear-engined V-8 with all-enveloping streamlined body.

HANS LEDWINKA

HANS LEDWINKA, AUSTRIAN designer of Czech automobiles, eventually achieved the status of motoring prophet with considerable honour in his own country—largely because he managed to live and create through seven decades of wheeled history, beginning in 1897.

The son of a barracks canteen manager, raised by and apprenticed to a machinist uncle, Hans devoted his instinct for materials and production techniques to cars, railway locomotives and—by preference—heavy trucks. His life-work was crowned by gold medals, government 'Orders: First Class' and an honorary doctorate of engineering. But, apart from the latter belated degree, bestowed on him in Vienna in 1944, this flood of honours was to come well after his pioneer designing days—and a six-year period of political imprisonment—were history. The homeland was strangely, if conventionally, indifferent to its loyal prophet for some fifty years.

Born on 4 February 1878 in Klosterneuburg, near Vienna, Ledwinka was a true subject of the Austro-Hungarian monarchy, speaking only German even when he was Tatra's design chief under the Czech republic. Through all the many Central-European frontier changes, he steadfastly considered himself to be an Austrian, turning down offers of both Czech and German citizenship. Few men have enjoyed such a long and fruitful design career and fewer still have devoted their productive years so totally to a single firm. The Imperial Nesselsdorf wagon factory became republican Tatra in 1923, and Ledwinka worked for both with equal facility.

Apart from brief forays into steam and Steyr—both staunch Austrian ventures—his entire output fell quite naturally into three main epochs, each played against a Moravian backdrop. All three periods were marked by his lifelong fascination with passenger cars and/or trucks, for Ledwinka was never a real sports car fan like his Danube empire contemporary, Porsche, though he was certainly more of an innovator. True, he *introduced* such features as overhead camshafts, hemispherical combustion chambers, integral engine/gearbox units, forced-flow air cooling, central-tube frames and 'autobahn' streamlining to passenger cars before his peers—and even made them viable—but he never claimed to have *invented* them. Many of these innovations also appeared on his trucks.

Ledwinka's talents were primarily dedicated to load capacity, off-road manners and utility of trucks, and to cars for people of modest means. He was one of those rare pioneer automobile designers with

virtually no personal competition record and still less desire to commit
a factory team. Yet, ironically enough, modern Tatra fans are apt to
boast first of a 1925 Targa Florio class win, of embassy-black and rapid
V-8s turned into marathon winners, or perhaps of the 'better than
VW' peoples' car which German bureaucracy scuttled when the Reich
marched into Czechoslovakia.

But that anticipates our tale—a biography of remarkably orderly pro-
gression, and thus much like Ledwinka's cars.

*   *   *

As a youngster, Ledwinka worked for a time as a metal machinist in
his uncle's workshop and studied at Vienna's trade school before joining
a firm of railway coachmakers at Nesselsdorf, eastern Moravia, in 1897
in the middle of his nineteenth year.

The Nesselsdorf factory, founded half a century earlier by master
coachbuilder Schustala, had risen to international fame. By the 1880s,
it was producing 1,200 vehicles a year, had been designated as royal
coachbuilders by Friedrich Karl of Prussia, and was meeting steady orders
from Russia.

In 1881 the Neutitschein area acquired its first railroad and Schustala
agreed to build rolling-stock for it. An initial order for fifteen open
freight cars was followed by one for 120 flat cars before Hugo Fischer
arrived to head the department. By the mid-nineties, the Schustala sons
had sold out and Fischer became sole director of what had then become
a predominantly flanged-wheel business.

Fischer expanded into automobiles about the same time as his great
friend Löhner, but whereas Löhner had begun with a French Lefebre-
Fessard twin, which was not successful, Fischer leaned to a horizontal
Benz, encouraged by Theodore von Liebig, a pioneering Austrian
gentleman racer with a strong addiction to Germanic cars dating back
to 1894. In the spring of 1897 the Nesselsdorf factory bought a Benz to
study and began the manufacture of their first car, called Präsident, soon
afterwards.

Ledwinka had joined them in September 1897 at Fisher's instigation,
specifically to design railway coaches, but he soon took a hand in the
Präsident, which was growing in a corner of the shop and already showing considerable improvement upon the Benz. He played no very important part in this first Nesselsdorf car, which was assembled under the
direction of master mechanic Svitak, but he could not help noticing its
flaws. Svitak, incidentally, was given a measure of inventive credit by
Communist writers in the 1950s, but Hans's son, Erich, is emphatic in
stating that Svitak simply oversaw the assembly of purchased parts.

Page 117 (above) Hans Ledwinka, a photograph taken during the 1930s when he was at the height of his design fame; (below) the first 'Präsident' being assembled in a corner of the works at Nesselsdorf

Page 118 (*above*) The chassis of a Nesselsdorf Type U, Ledwinka's last design for the factory prior to his departure in 1916. The engine was a six-cylinder unit of 65hp and the car had four-wheel brakes; (*below*) the 1924 Steyr tourer with overhead-camshaft engine

Page 119 (above) The Tatra twin in process of gathering laurels in the 1924 Alpine Trial; (centre) an overhead view of the backbone chassis on the 1924 Tatra twin; (below) the imposing chassis of the water-cooled V–12 Tatra Type 80 of 1927/28

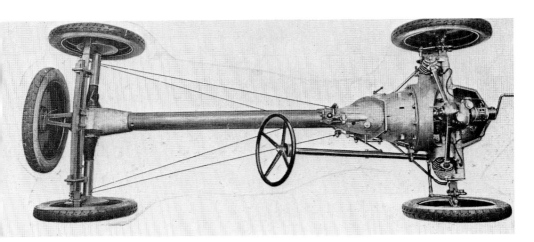

Page 120 (above) A Type 87 of 1937, the second generation of Ledwinka's rear-engined, air-cooled V-8 streamliners. With a 2·9 litre engine, the Type 87 was credited with a speed of 100mph and achieved 20·4mpg in observed tests; (below) rear suspension and engine unit of the 1935 V–8 Tatra

Nesseldsorf's two engineers, Rumpler and Sage, directed design and young Hans, already known for his skill in producing rapid, precise mechanical drawings of his own, was chiefly employed in clarifying doubtful points in the working drawings. The Präsident retained the basic shape of the Benz together with the Victoria's two-speed belt drive, and gear changing was effected by pushing the steering column forward (for hills) and pulling back for high gear. Neutral lay in the middle. This vehicle may also have had the world's first front bumper, and though it fell off within the first ten miles of the car's initial test run to Vienna, it did at least survive long enough to form the basis for several subsequent patent fights.

Alfred Neubauer, later to become world-famous as manager of the Mercedes racing team, recalls that Ledwinka used to pace the first Präsident on his bicycle. Later on, when Hans was courting his first wife, he used to visit her by car at nearby Neutitschein twice weekly, and Neubauer was not slow to take this opportunity of examining any new model he had failed to see when accompanying his cabinet-making father on window deliveries to the Nesselsdorf plant.

In 1898, director Fischer sent the factory's first product to a Viennese celebration of the Emperor Franz Josef's fiftieth jubilee, with instructions that it should be thoroughly tested on the way! The gathering was proclaimed as a grand assembly of Austrian-built automobiles—four in all. One was the Marcus, then dated 1877, Löhner contributed two, one petrol-engined and the other electric-powered, and the fourth was the Präsident.

The Nesselsdorf car made the 204 mile journey to Vienna on 21-2 May, escorted by relays from local cycling clubs after Ledwinka had turned back. The journey took just fifteen minutes over twenty-four hours, of which fourteen and a half hours were spent in actual driving. The Benz which had been bought as the original pattern came along too, and both vehicles were subsequently presented to the Austrian Automobile Club which Fischer and his friend Löhner had helped to found. Later, that first model of the Präsident was to be returned to the factory and presented to the Czech republic in 1919. It stands in a Prague museum today.

With the car's durability proven to their satisfaction, the factory laid down ten more, this time dispensing with Benz parts in favour of factory-made pieces supplemented by some special castings from a firm in Vienna. Chains replaced belts for the final drive, there were four speeds independent of that shifting steering column, and pneumatic tyres were an optional extra.

The new gearbox on this improved model was directly responsible for Ledwinka's rise at Nesselsdorf. It had been designed by the firm's two engineers, Rumpler (later of aeronautical fame) and Sage, and while it was most ingenious it just did not happen to work. Both engineers were dispensed with—the common fate of Moravians who failed at their first attempt—and Hans Ledwinka was given his first solo task, to design a gearbox which *would* work.

This it certainly did, and well enough for Liebig driving 'Wien'—each car was individually named—to take a 'first' and a 'second' among gentlemen drivers, plus a 'third' on handicap, in a race meeting at Prater. Later, in March 1900, another car won a hill-climb in France, so that Ledwinka's products were racing early when he was in no position to approve or disapprove.

By then, a Nesselsdorf cost 6,600 crowns with solid tyres and 7,200 'on air', and it was evident that the carriage-style body was not ideal for racing purposes. Prodded by Liebig, Ledwinka fitted a 12hp Mannheim engine into a new 'racer' in five weeks. This car, which did 57mph and ran second to a Daimler of much greater power, also now stands in the Prague museum—on pneumatic tyres.

In the spring of 1900 Nesselsdorf took one of the largest stands at the Austrian Automobile Club's salon and showed a bus and six passenger cars, including four-seaters with twin-cylinder engines of 6 and 9hp and four-speed gearboxes. The bus was powered with a French steam engine.

By 1901, Ledwinka, with two young assistants, was in full control of automobile activities and it was then that the carriage shape was abandoned and all effort concentrated on a new model built entirely in Nesselsdorf, with the engine slung beneath the chassis and a conventional steering wheel. This 'underfloor four' was so flexible that the car would sail up a steep hill in Vienna which was used for driving tests. The first examiner to ride in one promptly asked for the bonnet to be opened, gazed in awe at the huge cylindrical muffler—all that he could see—and exclaimed, 'It's incredible that a single-cylinder machine can climb such gradients!'

It soon became apparent to the factory that there was no future in manufacturing cars in small numbers and that production would have to be considerably increased if it was to become a paying proposition. This brought another problem in its train, for more cars meant more testing on local roads, much to the disapproval of officialdom in nearby and more self-important towns such as Neutitschein, the county capital. In those days, speed in Moravian towns was restricted to that of a horse's gentle trot, a rather faster gait was permissible outside built-up areas but only on 'wide, straight and relatively empty roads'. On curves and during market days speed had to be held to a pedestrian's pace. In September 1901, the authorities were reporting 'many complaints' about the Nesselsdorf cars and wagon drivers were warned to keep a tight rein on their horses when one was seen, or even to dismount and hold the bridle. Car drivers themselves were not permitted to leave their vehicle while the engine was running, nor to allow an 'unqualified' man to start it.

Such were the local restrictions faced by all the basic Nesselsdorf models built between 1900 and 1905: the Model 'A' with an 8hp opposed twin in the centre of the chassis, the Model 'B' of 12hp with a six- or eight-seat brake body, and the Model 'C', a 24hp opposed four. Both the 'A' and 'B' models came directly from Ledwinka's drawing-board but he had no part in the design of the Model 'C', which was strongly criticised in March 1903 in a motoring paper—in which, incidentally, Nesselsdorf had not advertised for the previous two years!

On 1 September 1902 Ledwinka left for the first of his two brief absences from the Nesselsdorf firm. One reason was that he had become increasingly interested in steam cars and the flexibility of power which they offered. Another was his constant disagreement with the foreman, Svitak, on various aspects of design, and a third was his realisation that cars were regarded only as an expensive sideline by the firm's directors.

Five years to the day from his first arrival at Nesselsdorf Ledwinka joined the firm of Alexander Friedmann in Vienna where he worked under Knoller, a first-rate theorist and later professor at the Vienna technical college. Among their first joint achievements was to fit one of the Friedmann steam cars with front-wheel brakes. During an early test, these locked with predictably massive understeer, the flaw being the system of pedal operation for the front brakes and a hand lever for the rear pair. The Austrian motor press were quick to point out that any car with front brakes must inevitably turn turtle. Ledwinka was unimpressed and filed the system for future development.

The steam car was also being built under licence by Weyhr & Richmond of Paris and this gave young Hans his first taste of travel and of French cooking, both of which became lifelong enjoyments. He stayed in Paris for nine months in 1905 and then returned, on 1 December, not to Vienna but to Nesselsdorf. Svitak had retired and Ledwinka, then just twenty-seven years of age, was appointed director of the automobile section, a somewhat hollow honour since the four-wheel side of the business was virtually moribund. In his absence the board, realising that they were lagging behind the times, had given two engineers, Kronfeld and Lang, separate briefs to design new, vertical, front-mounted engines in original chassis. Lang had also built an opposed twin for quarry locomotives. The net result had been three fiascos in a row and Ledwinka's first task was to try and make something of the two cars.

One car boasted a 5 litre T-head four-cylinder engine with chain drive in a chassis with a bare weight of two tons. The power output was hopelessly inadequate and Kronfeld, its designer, soon left. Lang had opted for shaft drive on his 4 litre four with paired cylinders and magneto ignition. His power curve was as disastrous as Kronfeld's, his chassis was not worth development and he, too, left the firm—though there was not much of it to leave by then. Its directors were all thoroughly fed up with designers in general but Ledwinka's faith in the future was unshaken and, with the aid of a young helper, he embarked on a dawn-to-dusk recovery effort. The car that resulted became the Model 'S' and was far enough ahead of its time to put Nesselsdorf back into the automobile business to stay. It was also Hans Ledwinka's first complete design concept.

The original Model 'S', which Ledwinka had begun designing in 1905, had four cylinders cast in pairs because of foundry limitations, with hemispherical combustion chambers—some fifteen years before Ricardo used them—a single overhead camshaft and valves in screw-in cages at an angle of 45 degrees to the cylinder axis. The crankshaft was inserted into a barrel crankcase where it rode on three roller mains, while a vertical shaft transmitted cam drive and a belt from a pulley on the camshaft-nose turned the fan. A bore and stroke of 90×130mm gave a capacity of 3·3 litres and produced 30hp at 2,200rpm.

Antonin Klicka, who was employed to do Model 'S' drawings, vividly remembers how quickly Ledwinka used to sketch out his ideas in the glass box they shared in the middle of the assembly hall. And as the 'boss' showed little interest in workroom comforts, the design office had to get along with only two chairs—one apiece! Ledwinka, he recalls, also liked to talk to himself when at work. On one occasion he roughed out a crankcase access panel using his own hand for scale. Then second thoughts set in. Studying the sketch again, he said, 'Look at you, Hansi! Other men have larger hands. Make it bigger!'

Work on the Model 'S' moved rapidly out of the drawing-board stage, 'and there was no need to add or subtract at any point from his first sketches or dimensions,' Klicka notes. 'Then, when something didn't please him in the shop, he never said it was bad work but simply started from a completely different viewpoint. Perhaps he'd suggest we might try it "this way", always with a lightning drawing to illustrate what he had in mind.'

Along with this advanced—almost racy—engine, Ledwinka designed an entirely new gearbox, using known features as always, but turning them into a workable production combination. He was already showing that preference for attaching engine to gearbox which was to be carried through most subsequent Tatra cars and trucks. The common 'crash box' of that era required skilful handling, so Ledwinka substituted radial engagement of an entire toothed flank for the sliding axial mesh. This so-called 'bell' box was mated to an oil-bath, cast-iron cone clutch, an idea taken from American machine tools. Only top gear was engaged by sliding the gear wheel axially to meet its pinion.

The managing director, Fischer, had been in Vienna while the new model was being rushed through and did not see the first car until it had already been tested. While he and his board were debating its chances of enabling them to stay in the car business, ten more were being quietly built in the works and, what is more, had already been sold in advance. With still more orders actually in hand, Nesselsdorf decided that there might be a future after all for the automobile!

Realising that the continuance of the car division—and, incidentally, his own job—depended very much on him, Ledwinka was spurred to further effort and by 1909 he had applied a Friedmann lesson by fitting rod-operated four-wheel brakes, and all four cylinders of the Model 'T' were being cast *en bloc*.

No doubt much of his early success was due to his personal supervision of the shop work on each prototype. Klicka remembers him as having 'a good heart but a firm hand with any workmen who malingered'. And already he was beginning to assemble his own team around him, men who were to remain notably loyal to him through successive changes of government.

He was certainly more high-handed with his management than with his men, and when the board proved unwilling to commit themselves to another design, Ledwinka, encouraged by a friendly industrialist-cum-customer, went ahead and built a six-cylinder car just the same. The industrialist, Fritz Hückel, of the Austrian hat 'empire', wanted a larger engine and Ledwinka had 'just happened' to design the Model 'S' so that another pair of cylinders could easily be added—once the necessary parts had been machined outside the factory to blueprints he also 'just happened' to have on hand!

Hückel—whose daughter later married Ledwinka's son, Erich—not only sold the firm's directors on the completed 'six' but also won a sizeable bet as to its flexibility by driving from Vienna to Stettin in seven minutes under the prescribed twenty-four hours with the car locked in fourth gear. The gentle take-up of the oil-bath clutch helped considerably in the achievements of this feat.

This S-50, as the new 'six' was called, earned its small niche in racing

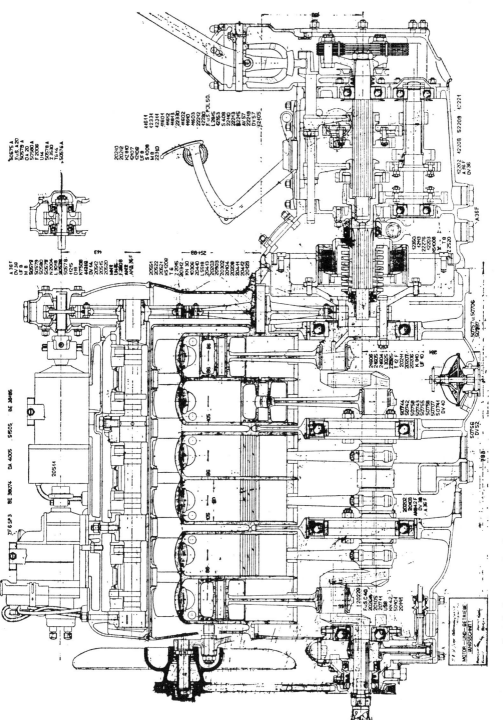

Ledwinka's six-cylinder ohc engine, in unit with the gearbox, designed for Steyr in the early twenties

history when Ledwinka himself drove one, as did Hückel, in the Alpine Trial of 1911. Neither collected a penalty point but though his chauffeur later raced the car, as did many purchasers who relied upon its reliability rather than its speed, Ledwinka was never a believer that racing did much to improve his breed of car.

Turning instead to the possibilities of goods transport, he fitted small truck bodies to the 'S' and 'T' chassis, with little modification other than a change of final drive ratios. The last notable design in this series was the Model 'U', which was first marketed with four cylinders, around 1914-15, and later appeared in 65hp six-cylinder form with four-wheel brakes.

Production climbed steadily with the war and Ledwinka asked for a new building to be put up, which the board approved. But when Köbel, a new manager, used the money to build a railway-coach workshop instead, Ledwinka accepted an offer to become head designer of a new automobile division then opening at the Steyr arms works. Unable to leave Nesselsdorf at once, he first worked on Steyr's new 'six' at home, visiting the Austrian plant every few weeks until he was able to move his family back to the 'homeland' in 1916. He also took with him several good engineers and master mechanics from the team he had built up at Nesselsdorf, and it was his practice to visit them and tour the entire plant every morning to check work in progress and tool purchases. He also personally supervised and controlled every facet of production.

The seventy-five-year-old Steyr ironworks at the junction of the Enns and Steyr rivers had hoped to produce its first automobile by 1916, but even a Ledwinka could not achieve that miracle and it was another three years before it made its appearance. So it is hardly surprising that the 1920 Steyr 12/40 six, an elegant 3·3 litre, had such up-to-date features as a barrel crankcase, overhead camshaft and engine integral with gearbox. The 40hp engine used a three-piece crankshaft flanged together with a ball race at each joint. It also embodied Ledwinka's screw-in cages for the overhead valves. (See drawings, pp 126 and 128.) Officially, power peaked at 2,400rpm, but there are those who still insist that the safe maximum was 4,000 or even 5,000. A four-speed gate change and laminated clutch broke new Ledwinka ground but he dropped four-wheel brakes for this first Steyr, though he reinstated them on the second model.

Some 6,000 assorted Steyrs had been built by 1925 so that the distinctive pointed radiator and model numbers like 'V' and 'VII' became familiar on Continental roads. The number 'IV' was assigned to a smaller car with a side-valve 'four' and 'bell' gearbox which, though it

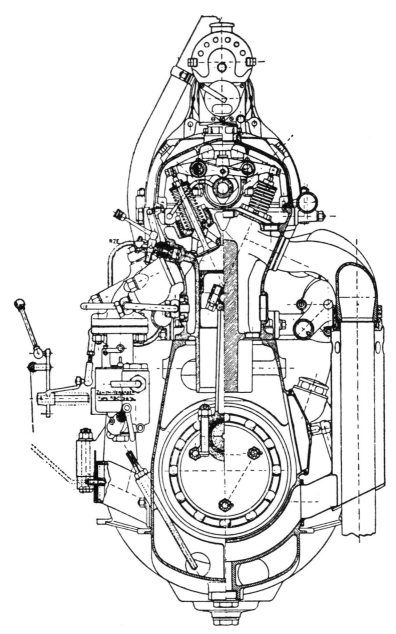

Cross-section of the 3·3 litre six-cylinder Steyr engine

never reached the production stage, is indicative of Ledwinka's early interest in cheap transport. Most sportsmen preferred the 4 litre, 90hp Type VI, or even the 145hp Klausen (named after an Alpine pass), and when Porsche took over at Steyr in 1927 the designs he brought up to date were all Ledwinka's.

Ledwinka also built Steyr trucks, capable of carrying loads up to 2½ tons, around his six-cylinder engine, so laying the groundwork for Austrian heavy vehicle construction and reaping the kind of bonus which particularly appealed to him. The most stable market of those early twenties, however, was obviously for luxury cars, but Ledwinka none the less retained his faith in the need for small cars for Everyman as well. Many high-performance models were available at that time but relatively few cars designed from the road up with comfort and reliability just as firmly in mind.

Characteristically, he went ahead with his ideas on these lines, designing at home a simple, rugged, roadable small car while the Steyr directors vacillated and asked him 'Who wants small cars these days?' and why he was building them a 'six' when the French and Mercedes were offering 'fours'. Their indecision, coupled with repeated overtures from Czech directors, who had even built a new works to meet Ledwinka's main objection to the old firm, finally lured him back to what had now become Czechoslovakia, in effect a 'foreign' republic, but no sooner was he installed at the newly-named Ringhoffer wagon factory—the town had become Koprivnice instead of Nesselsdorf, too—than Steyr emissaries arrived bearing blank contracts.

Not being particularly interested in money, according to his son Erich, and having no inclination to switch back again, Ledwinka came up with a compromise. He suggested that a central development office for both Ringhoffer and Steyr should be set up at Linz, in Austria, and that he should there design for both firms. This project fell through but Ledwinka did subsequently make many visits to Steyr to advise on work in progress, which explains certain similarities between his new Tatras and cars from the Austrian plant.

The problem of transition to a country where three out of four of his workpeople normally spoke Czech—a language Ledwinka never learned—he simplified by taking with him many Steyr people, both Nesselsdorf men and later adherents. As this engineering staff spoke more German than Czech, in roughly a reverse 3:1 ratio, all drawings were captioned in that language.

Ledwinka was convinced that a good small car could never simply be scaled down from some larger model, particularly in view of the appalling state of the roads in Bohemia and Moravia in the twenties. The car

H

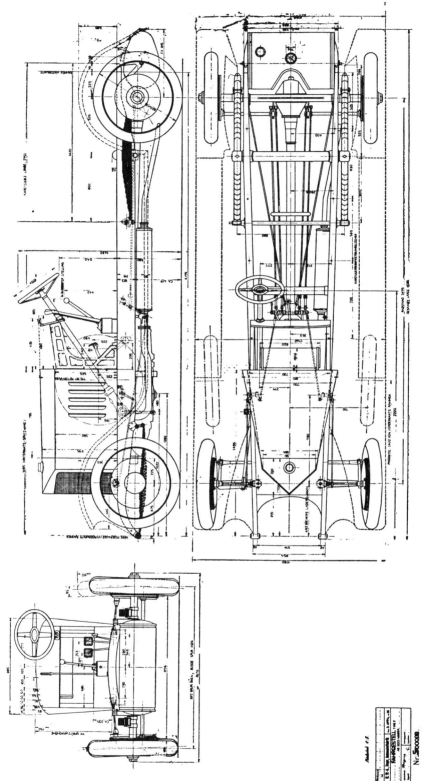

The Steyr Type VII of 1925

he had in mind, and the design he had first shown to Steyr, was a completely original one and since he had done all the work on it at home—and had dreamt up many of its features while recuperating in hospital after breaking an arm when his car had overturned near Steyr—he felt justified in taking the plans along with him to Czechoslovakia in 1921.

This design, Ledwinka's second major motoring revolution, was for a frameless (central tube) light car powered by a forced-air-cooled opposed-cylinder engine placed well forward over the front axle, with independent rear suspension by half shafts without universal joints. It became, of course, that famous first Tatra which Czechs called the 'Tatrachek' because of its diminutive size, which the Austrians nicknamed 'the tin dachshund', and the factory—more prosaically—called the Model 11. While no prototypes now survive, Klicka recalls that Ledwinka did all the detail drawings within a few days of returning to Tatra, where he was to remain until 1945.

At forty-four years of age Ledwinka now entered upon the most productive and certainly the best recognised stage of a long career. The T 11 was the cornerstone of that career, in chassis and suspension as well as engine though, as regards the latter, Julius Mackerle, the Czech specialist in air cooling, has pointed out that it was far from easy to 'sell' the system to the public. Ledwinka, however, was wholly committed to it, retaining the system right up to Model 87, though he was always well aware that it would continue to be acceptable only as long as he kept driver and noise apart. He retained forced-air-cooling, too, even for those massive diesel-engined trucks which are still produced at Tatra, and bearing witness to the soundness of his conviction are the many T 11 (and very similar T 12) cars still running in Czechoslovakia, often with half a million miles on the clock.

Power for his small car came from a 1·1 litre overhead-valve twin-cylinder engine of 12 and later 14hp. The cylinders were partially shrouded to control airflow from the flywheel, which doubled as a fan. The engine was attached directly to a normal gearbox connected to a one-piece drive shaft running down the 4·3in diameter backbone tube. This shaft ended in a spur-wheel differential and two spiral-tooth pinions engaging crown wheels of differing diameter but identical tooth count. These were fixed to the swinging half-shafts which moved around the pinions in a vertical plane, so dispensing with the need for universal joints. A later modification set one road wheel and its half-shaft slightly behind the other, allowing equal-size pairs of crown wheels and pinions. Tatra trucks, nearly half a century later, still retain this system. There were transverse springs at each end, the front one being attached to a light-alloy pan under the engine, the rear one to the top of the differen-

tial casing. Drive and braking thrust forces were transmitted to the central
tube by forked arms swinging with the half-shafts.

The very similarity of the design enabled the weight to be kept down
to 1,500lb in a car capable of 44mph, and Ledwinka's pioneering achievement consisted of fitting his air-cooling system, backbone chassis tube
and jointless axles into the one automobile. He never claimed to have
invented any of these features, but he *did* make them all work har-
moniously together in one motorcar.

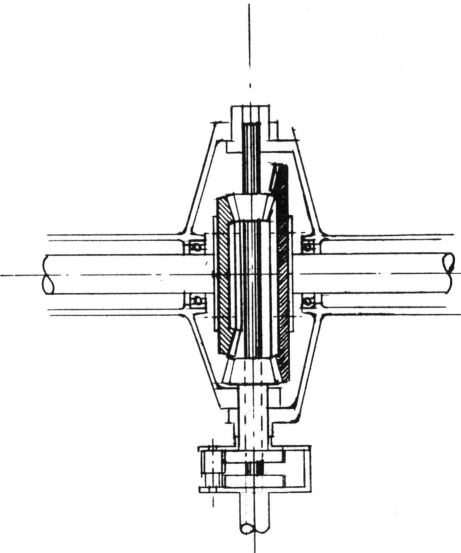

Early Tatra swing-axle design, with spur-wheel differential (below) driving two
concentric pinion shafts and the two crownwheels swinging with the axle shafts

Ledwinka has been called the inventor of the jointless swing axle, but a similar system was patented by Adler in 1903 and tried by La Buire, of Lyon, in 1909. An American had patented the central-tube chassis without producing any cars, while air cooling was familiar enough, though generally considered to be inefficient. Ledwinka's contribution was ducted flow.

His chief aim in replacing a perimeter frame by the central tube was torsional stiffness to allow lighter bodywork, and Mackerle recalls that Ledwinka badgered his body man daily, demanding new T 12 forms every two weeks. The original car had an open four-seater body but a saloon, coupé, 'Normandie', and cabriolet versions were soon added.

However busy the factory was, Ledwinka was insistent upon meeting deadlines, however much overtime might be involved. While following work in progress he would sign letters, or even dictate, while on the shop floor, and even when returning from a trip after midnight he was quite likely to drop in at the factory for a quick look round. He always looked after his staff well, often ordering sandwiches or drinks for them during overtime, and as the boss ignored late hours none of his engineers ever liked to remind him of the time—or their own burnt-up suppers waiting at home.

Ledwinka made a practice of driving the first few hundred yards in each brand-new prototype, after first examining the whole car, trying the seating and testing the controls. If he came back from such factory-yard sorties smiling, all was well, but 'all hell broke loose' (in Mackerle's words) if he found a real fault in assembly. Poor basic design was something else. He seldom became excited at that but would sit in his office for hours on end working out corrections. Once convinced that there was no simple solution, he was quite capable of scrapping an entire project—a fate which later befell the T 90, a passenger-carrying three-wheeler.

When his T 11 was a reality, Ledwinka loaded his wife and two sons into the first four-seater and covered nearly 6,000 miles of Austria, Italy, Switzerland and France. He drove all the way, though he usually preferred a chauffeur when space allowed, even though more than one had deposited him in a ditch. He liked to be driven fast.

By 1923, Ringhoffer had amalgamated with his wagon factory in Prague to form the Tatra company, named after some nearby mountains, and it was felt that a sporting cachet might help the new image. The year 1925 is usually said to have marked Tatra's sports début but, in fact, a T 11 had won its class on Solitude in 1924. Such cars, however, had been exceptions to the firm's earlier ban on competitive motoring, and were special versions with higher compression, dual inlet valves, roller bearings and larger carburettors. In 1925, however, a more serious effort was made. A pair of cars were fitted with mechanical four-wheel brakes, Rudge-Whitworth wheels and swing axles in front (at the behest of Fritz Hückel and Karl Sponor), thus placing them among the earliest all-independent racing machines. With an admitted 35hp and a maximum speed of nearly 85mph, these two cars finished first and second in the 1,100cc class over the 1925 Targa Florio course of 67 miles to the lap.

This was the only victory ever celebrated in a big way at the factory—perhaps because two directors were involved! Normal works successes were apt to be ignored, with the signal exception of the original Ecce Homo hill-climb at Starnberg when Ledwinka embraced the winner, Vermirovsky and said 'Pepek . . .' but was too overcome to continue. Previously, he had fought against racing the baby Tatra, fearing that it would not stand up to competition driving, but after this Ecce Homo win he changed his mind sufficiently to study lightweight racing bodies. Even so, long-distance endurance runs were the only form of competitive motoring in which he was ever really interested.

Another Vermirovsky triumph, in a 1925 production car with beam front axle, was to win the Leningrad–Tiflis–Moscow reliability trial over 3,300 miles of roads and steppes—at least 60 per cent being the latter—defeating seventy-eight entrants in both fuel consumption and reliability classes. Not long afterwards, one of these cars was to cross Africa from Alexandria to Cape Town while another would span the Australian desert.

Vermirovsky raced or rallied Tatras whenever the firm would let him, until he eventually retired to manage the Tatra museum. A typical Ledwinka assistant, he had joined the firm as an apprentice and one day found himself with another youth shovelling coal in a wine cellar. Unable to resist such largesse they drank a bottle, then broke it and sprinkled water around to make it look like an accident. Their deception lasted less than a day, and Vermirovsky still remembers Herr Direktor Ledwinka's carpeting and the unanswerable question: 'Did you really think I was so stupid as to believe in a broken wine bottle with its cork removed?'

The Ledwinka who could stay a jump ahead of his staff was equally at home with Baron Ringhoffer—a proprietor, at last, who left his designers strictly alone and was interested only in results. This is the chief reason Erich Ledwinka cites for his father never building a car called the Ledwinka. Nor did he ever seriously think of setting up his own design office.

Ledwinka was never tied to the time-clock at Tatra; he would go home for a mid-afternoon rest if he felt like it and generally did very much as he pleased. In short, it *was* his own company so far as design and production were concerned, and he had no time or liking for financial accounts anyway. Erich, who worked alongside his father from T 77 days onwards—'with no special favours whatsoever, even though he was pleased to have another designer in the family'—feels that Ledwinka's strength lay largely in an ability to pinpoint the solution to any problem at first glance, no matter how many possibilities might be offered.

Once the T 11–12 twins were established, the next project was an opposed four-cylinder car based on the same chassis and engine concepts. Known as the Type 30 with the 1·7 litre engine, or the T 52 in 1·9 litre form, these cars demonstrated to the market of the late twenties that comfort need not be synonymous with weight. Small Tatras performed admirably over potholes. But most popular of all was surely the T 57, built from 1930 to the Second World War, with independent front suspension and a four-cylinder 1·16 litre engine of 20hp.

The high status of these Czech cars led Röhr of Darmstadt and Stoewer in Stettin to build versions under licence which were called 'Junior' and 'Grief' respectively. Even Daimler-Benz seriously contemplated arrangements with Tatra in the twenties when looking for a small car for their production series. And, almost as a sideline, Tatra also produced a prestige line of 6–7 seaters cars with water-cooled engines, like the T 70 six and the V-12 Type 80 of 1927–8. Both featured engine-gearbox units, central chassis tubes and swing-axle suspension.

Looking ahead as usual, and realising the noise limitations of air cooling, Ledwinka now envisaged an engine hung on the rear of his famous chassis tube, where noise would be carried away. So it was that the forerunner of the famous V-8 Tatra limousines carried its air-cooled twin-cylinder engine in the tail. Demand for the T 57 shelved the project at the time, though it reputedly gave several ideas to Porsche, whom incidentally, Ledwinka met several times, though not as often as one might have expected of two men of similar interests who were usually working within a day's drive of one another.

Moving the engine rearwards was mainly for decibel reasons, but Erich Ledwinka admits that it proved doubly acceptable on streamlining grounds. An elongated tail and short nose were well attuned to contemporary thinking. 'Bodywork and chassis design were all one to us,' he says, 'so my father often took a hand in what is now called "styling", adding a line here, altering another there and making the final decisions.'

Aerodynamic shapes for automobiles were by no means new, but most attempts had foundered on the problem of engine location or perimeter frames. Replacing the box with a central tube, forked at the back to enclose its engine, helped the stylists no end, and Ledwinka even sheathed the underbody to improve the air flow. With such tricks he achieved an air flow or C_w value of only 0·212, ultimately giving the Type 87 a (claimed) 100mph top speed on 2·9 litres. By then he was using twin fans and aluminium cylinder heads, and applying many lessons learnt from the pioneering T 77. This earlier '77' had had a 3·4 litre V-8 engine behind its rear axles, sprung by a transverse leaf, whereas the '87' had outwardly-splayed quarter-elliptics.

Gear-change rods, cables, fuel lines and suchlike ran down the central tunnel while a battery and two spare wheels, as well as hand luggage, were carried in the nose. Main boot space was found above the rear suspension, the passengers riding between the axles—where Ledwinka always wanted them to be.

This T 77 was first unveiled before the Prague press on 5 March 1934, the '87' following in November 1937. The latest car weighed 4,020lb (20 per cent less than a '77') and gave a much softer ride. The engine, with chain drive to its overhead cams, produced 75hp. Ledwinka was chiefly concerned with reducing engine weight in the '87'—no bad idea considering its outrigger position. But the real paradox of the car lay in its exceptionally soft travel over unmade roads allied to stable high-speed cruising. In trials on the Padua *autostrada* during the autumn of 1937, a T 87 achieved 94·5mph (100mph was always questionable) and did 20·4 miles to the gallon. The Tatra 603 built today for prestige use by Czech diplomats—older V-8s are popular as taxis—is based entirely on his design and uses many of the same materials.

The '87' achieved another distinction among the world's automobiles during the war, thanks to Dr Todt, the German *autobahn* expert. When the Germans arrived in Czechoslovakia Tatra was ordered to stop all production forthwith—an order which Erich Ledwinka believes to have been not unprompted by luxury-car, water-cooled pressures from the Fatherland. Later, they were allowed to continue with trucks and military vehicles only until, finally, Todt, who particularly favoured big Tatras, went right to the top and said, 'This "87" is *the* autobahn car.' The result was that Tatra became just about the only company in the world to continue building purely passenger cars, without a break, throughout the war.

They were less successful in saving that first Germanic 'people's car' project, the Tatra 97. Experiments with rear-engined twins, fours and eights had been whittled down to the V-8 and a 'pusher' four, which reached limited production during, and even saw a brief revival after, the war, though it received little attention then. One '97' was, however, tested near Padua, alongside its big brother. Producing 40hp from 1,761cc, the 2,535lb car returned 73·9mph and 24·1mpg. All Tatra enthusiasts and many motoring historians believe it ran too well and was therefore stifled when Germany annexed Czechoslovakia.

Tatra had some eight to ten claims for patent infringements against the VW, or KdF as it was then called. These included ducted air cooling, engine position and gearbox layout, but when Porsche brought these up to the attention of headquarters he was told that the German construction front would settle all patent bothers and that he should go on building cars, using whatever sources he chose. By this time it was 1938 and the problem had become academic, though Ringhoffer's heirs did receive a large settlement after the war. Hans Ledwinka assisted with their case, not so much for personal gain as to get the record straight. Tatra, incidentally, had once lost a similar patent fight to Austro-Daimler over swing axles.

In view of the suppression of the T 97, it is ironical that Hitler had an early fondness for Tatras. When he visited the 1933 Berlin motor show, he went straight to Tatra's stand to tell Ledwinka how he had used a Tatra twin for a million kilometres of political motoring around Austria. Ledwinka had to explain the air-cooled, rear-engined V-8 to him in detail, and then return at ten that night to explain the car all over again when Hitler had more time to spare. Hitler's final comment was that any new German popular car must be just like a Tatra—air-cooled and robust.

By the time of the 1934 Berlin show, Hitler's party was expected only to visit good Teutonic stands, but again their leader spotted a Tatra sign and came over. Mumm, of the sales department, introduced Erich Ledwinka while Hans remained half-hidden in the foliage. Hitler sought him out to ask, 'What do you have for me that's new this year?' The best answer, apart from a '97' of course, would have been: 'Air-cooled trucks with independent suspension. They'll go anywhere.' And still do. Four- and six-wheel load-carriers from the thirties run strongly today and the design is basic to all current Tatra heavy equipment.

Soon after his T 11 came out, Ledwinka extended the central tube and jointless swing axles system to trucks. The first was designated T 13, a one-litre twin from 1923-4. The initial six-wheeler came in 1925 and a 500cc three-wheel delivery van with swing axles in 1926. Prague buses featured swing axles in front before Tatra built a truck named 'Rumania' with a 4 litre air-cooled V-8 engine and six-wheel drive. Their heaviest, the T 111, could carry up to 12 tons, move at better than 40mph or climb a 50 per cent grade (60 per cent in reverse).

Ledwinka switched to air-cooled diesels in 1933, using the same 'add-a-cylinder' technique he had adopted in stretching the Nesselsdorfs' to a six. His truck engines had 110 × 130mm cylinders of 1·25 litres each and a built-up crankshaft allowing four-, six-, eight-, twelve- and eventually eighteen-cylinder engines. Power ranged from 70 to well over 210bhp. His 'eight' and 'twelve' were V engines, the 'eighteen' was a W. All featured axial blowers on the nose of each row, with direct injection and swirl pattern pistons.

When, much later, Ledwinka settled in Munich, he tried hard to sell swing-axle trucks to German industry despite their extra cost. He was also single-minded enough to mention them again at length in a Deutsche Museum speech in 1965, at the same time donating a T 87 to their collection. In fact, he never ceased to wonder why heavy trucks, doing far more road damage than cars, couldn't be softened with independent suspension. Alfred Neubauer remembers his boyhood idol becoming so excited that his high voice squeaked. Banging the table, he once cried: 'These beam-axle trucks—we're still riding around in carts.'

When one recalls that Ledwinka was originally engaged as a railway specialist, it is hardly surprising that many rail vehicles also came from his drawing-board. He turned out various self-propelled trams and short-haul railroad coaches, including a 'tower car' for the Czech state railroad of the thirties, in which the driver sat in a pergola amidships. This vehicle, which could run it in either direction, had a water-cooled diesel to soothe passengers' ears.

Ledwinka's next departure was a four-axle, high-speed, self-propelled coach with seventy-two seats, a buffet, two driver-posts and a pair of 175hp petrol engines feeding electro-mechanical gearing. This used a torque converter up to maximum speed whereupon the electrical system cut out. Drive then became mechanical and direct. The Slovakian 'Arrow'—as it was called—was streamlined to do 80mph, and reduced the Prague–Pressburg trip from 7 to 4¾ hours. Some were still in service on the Prague–Nurnberg line after the war.

Ledwinka directed all phases of Tatra production, which meant that he dealt with workers from every branch. A delegation from the rail division once came in to point out that motorcar builders were allowed to borrow cars and use them on weekend holidays to the mountains. They wanted the same privilege. Ledwinka studied them from behind his desk for a moment and then said: 'Gentlemen, I have no objection. Take a railway coach and go wherever you like.'

*   *   *

With so many design and production problems constantly on hand, it would have been strange indeed if Ledwinka had found time for politics.

But though he thus lacked the contacts needed to save his T 97, for instance, his trucks that survived were able to survive on merit alone in the face of German water-cooled opposition. He accepted an honorary doctorate in 1944 but turned down every other German medal, preferring to concentrate on his vehicles. Still, as Erich notes wryly, his father *was* a director at Tatra and had a share in the company.

When Communism moved into Czechoslovakia, Ledwinka was charged with almost every crime imaginable, arraigned three separate times and finally gaoled for six years on a charge of building trucks for the German army. He had been acquitted twice but each time word came from Prague to try again, ignoring over 1,200 witnesses who testified without exception that he was never anti-Czech, despite his Austrian passport.

When the government relented and released him in 1951, they suggested that Hans should remain and run Tatra. He fumed, 'Now you say that. Six years ago you should have offered,' and moved to Austria, where Erich was heading the Steyr-Puch design office, thence to Munich. He even refused an invitation to the Tatra anniversary in 1967 because the government would not release his entire confiscated capital.

When nearly seventy-five years of age and hailed by the German motoring press as Germany's most talented designer and a man happy to remain in the background while others took credit for his work, Ledwinka was still looking for new vehicles to design. He advised Perkins and Deutz on truck engines, even suggesting water cooling for a diesel bus engine at Deutz though the engineering world knew him as the 'father of air cooling'. Comfort always took precedence over pure engineering theory.

Sketching at home towards the end of his life, Hans evolved a pet project—a real city car based on the usual parameters of minimum exterior mass and new power train modes. He arrived at a three-seater egg, with the driver seated centrally in front while two passengers occupied the rear corners with their legs on either side of the pilot. That design was never realised, but not even Ledwinka could hope to build every idea that had occurred to him in seven decades.

Mackerle once found an archive design applying the stageless electro-mechanical rail carriage transmission to a Tatra 87. It was later forgotten, along with an air-cooled radial car engine which did reach the development stage and a pneumatic Tatra truck suspension which was actually tested. A propellor-driven sled also achieved prototype status.

Hans Ledwinka has often been called an individualist and nearly as often a genius. Though he received many awards in the fifties he was never mentioned in German reference works of any era, and the only biography of the man who was perhaps the last contemporary of Benz and Daimler was uncompleted when its author was killed in an automobile accident.

Ledwinka scored innumerable public and private triumphs with designs which the 'experts' pronounced unworkable, and was always willing to stake his reputation on any idea, however wild, once he was convinced that it would work. It may well be that his chief talent was his ability to throw away the ones that wouldn't before peers or public ever saw them.

In his role of motoring prophet, Ledwinka enjoyed three major epochs. The first embraced hemispherical combustion chambers, overhead camshafts, engine-gearbox modules and his original 'six', followed by heavy trucks with bevel-wheel drive and hub transmissions.

The second phase included Steyr's swing-axle cars—the sporting age of Hans Ledwinka, if you will. Innovation (and directorial support) were sharply reduced while production output ruled, and Steyr built on these foundations for some years.

In the third, and most fruitful, stage Ledwinka brought forth the non-whip frame, air-cooled engines with ducted flow in sizes from a 15hp twin to the eighteen-cylinder W diesel, all-independent suspension with jointless half shafts and marketable streamlining.

Disregarding railroad efforts, steam cars, automatic transmissions and city-car dreams, this constitutes high achievement and a most impressive corpus of work, even for a man who remained facile in design for nearly seven decades and who headed a major factory for half a century.

Hans Ledwinka died in Munich on 2 March 1967, shortly after his eighty-ninth birthday. He was one of (if not *the*) last of the true automobile pioneers, though no *marque* ever bore his name.

5 MARC BIRKIGT

by MICHAEL SEDGWICK and JOSÉ MANUEL RODRIGUEZ DE LA VIÑA

MARC BIRKIGT (1878–1953). Swiss: Designer-constructor of Hispano-Suiza cars, manufactured in Spain and France between 1904 and 1939; and of Hispano-Suiza aero-engines used in Allied military aircraft in both world wars.

MARC BIRKIGT

ON THE BULKHEAD of every Hispano-Suiza car to leave the Barcelona factory was a small plate bearing the words: *Sistema M. Birkigt Patentado.* Barcelona alone turned out 19,000 vehicles between 1904 and 1943, and to this impressive total must be added another 3,000 French-built cars from Bois-Colombes, not to mention countless aero-motors and 20mm cannon. The name of Hispano-Suiza is a legend on land and in the air, but the genius behind it, Marc Birkigt, remains a shadowy figure. He was as much 'Mr Hispano' as Cecil Kimber was 'Mr MG', and the latter half of the hallowed name commemorates his nationality, but he identified himself so closely with his brain-children that a separation is hardly possible.

Other designers were peripatetic. Porsche ended up as an independent with a bureau of his own, but before this he had been associated with Lohner, Austro-Daimler, Mercedes and Steyr. Ettore Bugatti was the uncrowned emperor of Molsheim from 1909 onwards, but he, too, had been a migrant, working for Prinetti *e* Stucchi, de Dietrich, Mathis and Deutz. Delage and Voisin headed their own concerns, yet Birkigt, though never the boss in the commercial sense, spent almost his whole working life with one factory. His earliest cars were La Cuadras and then Castros, but one can trace a continuity of design from the four-cylinder Castro of 1904 to the 30CV K6 Hispano-Suiza of 1934. Sometimes he worked in Barcelona, and sometimes in Paris, but always for the same master.

Birkigt has been compared with Henry Royce, and indeed the similarities extend beyond the double-barrelled names of their respective *marques.* Both served their apprenticeships in electricity and railways, though the railways were of vastly different types. If at their respective zeniths the Rolls-Royce and the Hispano-Suiza represented opposing approaches to luxury-car design, there were, none the less, parallels in their evolution, for both achieved fame as four-cylinder sporting machines of relatively modest price. Had Royce's TT-type 'Light Twenty', been developed, it might well have occupied the niche later claimed by Birkigt's 'Alfonso'. Both designers produced the ultimate in cars, though in different decades: the 32CV Hispano-Suiza of 1919 was the logical successor to the Rolls-Royce 'Silver Ghost' of 1906. More important, Royce and Birkigt made their greatest contributions in the field of the aero-engine: the outstanding liquid-cooled unit of World War I was the V-8 Hispano-Suiza, just as the 'Merlin', derived from Henry Royce's Schneider Trophy R-type, tipped the balance in favour of the Allies in 1940.

Nor does the parallel end here. Both the Englishman and the Swiss were dedicated to their work—hobbies meant nothing to them, though Birkigt was far more of a family man, and one whose courteous demeanour and diplomatic manner enabled him to dispense with a Claude Johnson as a channel of communication with others less irrevocably wedded to the drawing-board and the workshop. Both were fortunate in that they made the right contacts at the right time. Nobody in England was better equipped to promote a new car than the well-born and restless Charles Stewart Rolls, with his one-track mind; but in the more leisured tempo of pre-revolutionary Spain his ideal counterpart was that motoring monarch, Alfonso XIII. The two pairs of 'partners' were ideally matched. Rolls and Royce were monomaniacs of different species, whereas Birkigt, who appreciated good living and the pleasures of the chase, was entirely at home with the sovereign of his adopted country. Spain, and not Switzerland or France, came first in his affections. Finally, by an odd quirk, both Royce and Birkigt started their driving career on the same breed—Decauville.

Marc Birkigt was born in Geneva on 8 March 1878. Switzerland, with its tradition of precision engineering and horology, has bred many a great designer, and one French authority has summarised Birkigt's first aero-engine as *une merveille d'horologerie*. In view of this, one may well wonder why the Helvetic Confederacy has produced hardly any motor-cars of note.

The reasons for this are not entirely geographical, though the country's mountainous contours were daunting enough to pioneer motorists. But if feeble, constant-speed engines and crude brakes were allergic to the passes of the Alps, the Swiss were equally allergic to the automobile. Nobody could really blame the authorities for neutralising their leg of the Paris–Vienna route in 1902, but even then one canton (Grisons) imposed a total ban on motor vehicles, and Swiss speed restrictions outstripped the worst efforts of over-zealous Dogberries in Britain. On anything defined as a 'mountain road' forward movement was actively discouraged, and the Calvinistic *loi dominicale* ruled out Sunday joy-rides —in Berne the only permitted exceptions to this rule were doctors and veterinary surgeons. Worse still, such legislation still obtained in the early 1920s, and British tourists were warned against taking their cars to Switzerland lest they be stoned by the peasantry. Thus it is understandable to find Louis Chevrolet, Ernest Henry and Georges Roesch seeking their fortunes abroad.

Not that such a situation was relevant in 1899, when the young Birkigt took the road to Barcelona. He had been orphaned at the age of eleven, but his maternal grandmother, with whom he made his home, sent him to the famous Ecole de Mecanique, whence he graduated with distinction in engineering and physics before undergoing his statutory period of military service.

This was to have an important effect on his subsequent career, for he served his time with the artillery, acquiring first-hand knowledge not only of guns and their mechanisms, but also of the qualities of the different steels used in ordnance. He was one of the few car designers to apply himself seriously to armaments, and his ability as a metallurgist is shown by his adoption of nitrided liners in the 1920s. His military experience also helped his relations with War Offices in France and Spain alike; the armed forces were to be his best customers.

A civilian once more, he moved on to Paris, where he met a young officer of the Spanish Mercantile Marine, Domingo Tamaro. Tamaro suggested to his new friend that there were many promising openings for his talents across the border in Spain.

He was proved right, though Birkigt took some time to settle down. His first job involved the construction of some water towers for the city's Compania de Aguas. Then he was employed on a funicular railway being built between Barcelona itself and the docks—a 'natural' for a Swiss brought up on cables and associated problems of stress. He might well have followed this particular bent, but meanwhile his friend Tamaro had gone to work with Emilio La Cuadra, a manufacturer of batteries and electric vehicles. Further, the battery department was under the direction of another Swiss, Velino, so Birkigt was easily persuaded to change his allegiance.

La Cuadra, an engineer major in the Spanish Army, was also a man of some substance who had sold a power station he owned in Lerida, and devoted the profits from the sale to the establishment of his latest venture. He had been converted to the new locomotion after M Rasson had successfully ridden a De Dion tricycle from Paris to Madrid in eight days in 1897, triumphing over the crudities of his mount and the Pyrénéan winter alike. The Army, La Cuadra argued, could surely use motor vehicles.

It is not surprising that the major still thought in terms of electricity. The system avoided not only fumes and noise, but also the painful process of cog-swapping, while the batteries provided reliable illumination at night. At a time when long journeys were for maniacs alone, limitations of speed and range seemed unimportant. Tamaro and Birkigt, however, soon convinced their employer that the internal-combustion engine was a better long-term proposition, and their prentice effort was a 4½hp twin-cylinder affair with chain drive and steering-column change. This worked well enough, and half a dozen had been completed by the

I

end of 1901. Alas! Mechanical reliability is of no avail on its own, and
La Cuadra's money was running out. He found himself without
customers or backers, and though he took one of the cars on a trip to
Mallorca in quest of some additional finance, all he achieved was Spain's
first recorded automobile accident. The only casualties were a few
goats, but soon afterwards a strike dealt the *coup de grâce* to the infant
concern.

Not that all was lost, for among La Cuadra's creditors was a Señor
Castro who saw a future in the motorcar. Like many a dedicated man,
Birkigt could be most persuasive, and he seems to have talked Castro
into accepting responsibility for the firm's debts. Be that as it may, in
1902 the young Swiss had his first executive post as *chef de production* of
Castro's new acquisition. For the time being, existing designs were continued, but already Birkigt was ahead of his time and new cars with
shaft drive and mechanically-operated side-valves in a T-head made their
appearance. The largest model was a 2·2 litre four-cylinder which was
to form the basis of the celebrated 'Alfonso' Hispano-Suiza.

Spain, however, was stony soil for the motorcar, and by 1904 a committee of creditors was once again in charge. This time their chairman
was one Damien Mateu, and Mateu, like Castro before him, was prepared to back a Spanish-built automobile. Further, he recognised that
among the fearsome liabilities was one gilt-edged asset—Birkigt. The
Swiss was left in charge of design when the new Hispano-Suiza firm was
registered on 14 June 1904, and the name itself was a vote of confidence,
since it drew attention to the designer's nationality as well as that of its
manufacturers.

After two false starts, finances were now on a sound basis. Mateu, who
was to occupy the chair until his death in 1929, was an ironmaster and
an experienced businessman. His colleagues were equally dependable.
Seix y Barral was an editor and publisher and a keen driver, while Font-
cuberta, another Castro creditor, came from a well-known Barcelonese
family with wide financial interests. He bought the first Hispano-Suiza.
A proper sales organisation came into being with the appointment of
Francisco Abadal as the concern's Madrid agent, a role he fulfilled until
1912: Abadal also sat on the Hispano-Suiza board. By the end of the year
twenty-seven cars had been delivered, even if some of these were only
Castros assembled from parts still on hand at the time of the bankruptcy.

At this juncture Birkigt displayed a commercial wisdom scarcely to
be expected of a technical perfectionist. He took a step which would
have appalled Bugatti, and probably Henry Royce as well, by launching
a line of lorries. This was only commonsense, for these were needed at

home far more than were private cars. Spain is mountainous, and her railways were neither efficient nor profitable. As early as 1901, no fewer than fifty De Dion steam omnibuses were working in the country, and now Birkigt and Mateu offered something better suited to domestic conditions—a petrol vehicle of adequate power and modest weight, well-braked and easy to maintain. Further, Birkigt's technical expertise was at last backed by an ingenious policy aimed at attracting the customers, for Hispano-Suiza devised a species of hire-purchase in which shares in transport companies were accepted as part payment for new vehicles. This worked admirably, and even today one encounters names like Hispano-Leridana as reminders of this package deal.

Results were impressive. The customers, confronted with easy terms and excellent service, came back for more. Order-books were full, and foreign competition was kept at bay for many years. If no Hispano-Suizas were made after 1944, their tradition of high-grade commercial vehicles has been ably maintained by the Pegasos which succeeded them, and it is significant that more than half the old company's total output was made up of commercials.

It is conceivable that Birkigt might have made his name as a truck designer, but at this point a new name enters the story—His Most Catholic Majesty Alfonso XIII. The nineteen-year-old king was not only an early patron of the automobile, like Edward VII and Wilhelm II of Germany: he was an enthusiastic driver who sponsored the Catalan Cup Race. Alfonso was a true connoisseur, and owned a huge chain-driven Delahaye among other vehicles. He became acquainted with Birkigt's work during a state visit to Sagunto in the latter part of 1905. This involved a scenic drive to the top of the local mountain, and inevitably the price of an admirable vista was a procession of boiling radiators and protesting machinery. On all the cars save one, that is. The exception was, of course, a 20–24hp Hispano-Suiza, and within a year three of its sisters had joined the royal stud, the first of over thirty to grace the palace during the remaining years of the monarchy. Occasionally Alfonso was tempted by other marques—at the time of his abdication he owned a Model-J Duesenberg with coachwork by Hibbard & Darrin of Paris. But the Hispano remained his favourite.

Other heads of state followed his lead. The first of these was President Irigoyen of Argentina, who invariably sent his cars home to Barcelona for overhaul, while in the 1920s Queen Wilhelmina of the Netherlands owned a 32CV as well as her Spykers. Hispano-Suizas were, of course, the official carriages of later republican governments, and General Franco rode in a French-built 54CV during the civil war and thereafter, this one being supplanted in the early 1950s by one of the rare straight-eight

'Phanton IV' Rolls-Royces. Not a few Hispano-Suizas were sold to Indian princes.

In early days, of course, racing was almost a mandatory exercise for a manufacturer in quest of a reputation, and Birkigt was no exception. Further, his methods were a sure recipe for success. Charles Faroux once observed that any part of a Birkigt-designed engine which was subjected to stresses would not only be oversize, but twice as heavy as the norm, in spite of which, he added, 'a complete car will always come out a good five hundred kilos lighter than any of its direct competitors'. As early as 1902 Seix y Barral had won a race organised by the magazine *Los Deportes* on a Castro, but as yet Birkigt was content to submit his creations to long-distance runs. In 1906 a Hispano-Suiza was driven non-stop from Barcelona to Cadiz and back in twenty-six hours.

Hispano's great years on the circuits were, of course, between 1909 and 1912, the marque making its début in the Catalan Cup in the former year. One of the effects of a slump in the motor industry had been the voluntary abandonment of *grandes épreuves* by the major manufacturers, and thus the only serious races were staged for *voiturettes*. Hispano's *annus mirabilis* was 1910, when the long-stroke 2·6 litre T-head 'fours' of Zucarelli, Chassagne and Pilliverde took first, third and sixth places in the Coupe de l'*Auto* at Boulogne. This was not only the first foreign victory in the series sponsored by the French magazine; it marked the triumph of the multi-cylinder engine over a generation of freakish 'singles' and 'twins' with piston strokes of ridiculous length, and *ergo* the genesis of the small sports car as we now know it. Seen through modern eyes, this sounds like tendentious nonsense: the 3·6 litre 'Alfonso' Hispano-Suiza in production guise is quite a hefty piece of machinery, and its peak revs of 2,300 smack of cruelty when associated with a stroke-bore ratio of $2\frac{1}{4}$:1, but it must be remembered that even in the 'Alfonso's heyday the recognised method of catering for the enthusiast was to cram the largest engine possible into a liberally (and often unintelligently) drilled frame, and then add a vestigial body. Even some of the more modern designs were quite big: Porsche's 'Prince Henry' Austro-Daimler ran to 5·7 litres, and the cars it supplanted could boast as much as double this litreage. Birkigt did not make such mistakes: his chassis might be short on cross-bracing, but he ensured structural stiffness on even the earliest Hispano-Suizas by mounting the gearbox in unit with the engine, a principle that had become virtually universal by the later 1920s. The engines were also pressure-lubricated. Finally, these racing successes led to his migration (and that of Hispano-Suiza) across the border to Paris, then the hub of Europe's motor industry.

The cars were also raced in Britain, Belgium, Switzerland, and even

Russia, a 30hp model making the run from San Sebastian to St Petersburg in 1913 without even a change of tyres. Further, Birkigt's first great series-production model was a *fait accompli* by 1909, when the 15-45hp T-head sports model with monobloc cylinders acquired royal patronage and a new name. There is a Ruritanian charm about Queen Victoria Eugénie's birthday gift to her husband—a two-seater finished in white with gold lining and gold wire wheels.

In 1911 another name appears in the story—Birkigt's compatriot Ernest Henry, often regarded as the father of the modern competition engine, though in fact his twin ohc Peugeots were a direct crib from the work of Birkigt. The sixteen-valve layout, hemispherical combustion chambers, shaft-driven camshafts and dual ignition were all Birkigt-inspired. Henry, in fact, was really a skilled imitator who could combine Birkigt's ideas on structural strength with Bugatti's mastery of the high-speed engine. Be that as it may, the fact remains that in 1911 Birkigt, aided by Henry and Zucarelli (who was acting as chief tester), was working on a double-overhead-camshaft design to replace the old T-headers, and that December the first car was on the road. Within a month Alfonso XIII had sampled the prototype. He was favourably impressed—so much so that when Aritio from the Madrid agency suggested the name 'Prince of the Asturias' for the type, the King retorted that the car merited one name only, 'España'. In its original guise this new Hispano used a short-stroke (80 × 130mm, as against 80 × 180mm, for the 'Alfonso') 2·6 litre unit, though there was also a bigger 100 × 150mm version which Alfonso called the 'Super-España' ('Super-Hispano' in export markets). At this juncture, unfortunately, Henry and Zucarelli took off for France, carrying with them the designs of the new engine. These they sold to Peugeot, and within three months they had their twin ohc derivative running. An infuriated Birkigt took the case to court and won, but in the meantime he went to work on a variant with which he still hoped to beat the Peugeots at their own game in *voiturette* racing.

His solution was a supercharger driven by a twin-cylinder reciprocating pump *via* a forward extension of the crankshaft. Unfortunately, mechanical failures were to prevent the cars from running in the 1912 Coupe de l'*Auto*, but their development marked a turning-point in Birkigt's career. In 1912 Hispano's French agency at Levallois-Perret became an assembly plant, and 1914 saw the move to bigger premises at Bois-Colombes. Further, a chance encounter in a component store brought the designer into contact with Louis Massuger, who was to assume the role already filled in Spain by Seix y Barral.

For 1913, the organisers of the Coupe de l'*Auto* specifically barred

forced induction, but Birkigt went ahead with unblown developments
of the 1912 cars. These had very narrow frames, necessitating tandem
seats, while their high slit-like radiators contrasted sharply with the
'Alfonso's vast cooling surface: by this time, be it said, the old-type
sports car was making a name for itself with enthusiasts who appreciated
its modest thirst for fuel and tyres, its safe handling, and its 70–75mph
top speed. The racer's cramped proportions were alleged to have inspired its nickname of 'Sardine', but the truth is cruder and more mun-
dane. When the tandem-seaters ran in the Rabassada Races at Barcelona,
they left a strong whiff of castor oil in their wake, and some wag drew
the obvious and uncomplimentary parallel!

The Hispano-Suiza was internationally known by now, but Birkigt
was destined soon to follow the same path as Henry Royce, albeit with
more enthusiasm. Curiously, though, his aeronautical interests were inspired, not by France, but by neutral Spain. The Spaniards were quick
to realise that a European war would cut them off from their sources of
defensive weapons. As yet the country lacked an aircraft industry, and
in Madrid the Secretary of Defence, Amos Salvador, was pressing for
more money with which to develop aircraft and aero-engines, in which
policy he was backed by such senior officers as Kindelan, Ortiz Echague
and Vives, all men with flying experience. The result was a generous
subsidy, and this bred not only the ingenious Baradat-Esteve 'Torus' (an
anticipation of the Wankel) but also Birkigt's V-8, evolved with some
assistance from a Spanish Army engineer, Eduardo Barron. Barron, like
La Cuadra before him, had useful business interests and was director of
the military flying school at Cuatro Vientes aerodrome.

Once again, as in 1904 and 1911, Birkigt was ahead of his time, in
opting for a stationary engine at a time when the air-cooled rotary
was coming into fashion. Many of the earliest aero-motors had been
conversions of existing car units, and had combined excessive weight
with the inherent drag of unmodified plumbing. An engine capable (in
theory) of developing enough brake horses to lift a crude stick-and-
string biplane off the ground could weigh as much as 800lb. The rotary
circumvented this problem, but the presence of a large rotating mass in
the nose set an almost worse one of control, while the stationary radial
was not to catch on until the 1920s. Birkigt's water-cooled V-8 was, of
course, a direct descendant of his 1911 car engine. It weighed only 330lb,
or 210lb less than the 6·6 litre 32CV unit fitted to his post-war car,
and when type-tested in 1915 it gave a creditable 150bhp at 1,550rpm.
Further, V-type units lent themselves more easily to streamlining. Other
features handed on to the post-war cars included fully-enclosed valve
gear and screwed-in steel cylinder liners. Preliminary flight trials were

carried out at Madrid using a 'Flecha' airframe made by Barron: they were entirely successful.

News of these developments soon reached France, where the authorities were desperately seeking an answer to such German designs as the Mercedes, a close relative of that 4½ litre ohc unit that had so recently humiliated the Peugeots in the Grand Prix at Lyons. Already Marius Barbarou had left Delaunay-Belleville to work on captured German engines at the Lorraine-Dietrich factory. Captain Martinot-Lagarde of the Services de Fabrication d'Aviation Française was despatched to Spain. He was delighted by what he saw, and a Hispano engine was sent to the experimental station at Chalais-Meudon for further trials. Though a ten-hour test was deemed sufficient in those days, the Hispano ran happily for fifty hours at full power. It was promptly adopted by the Allies, the first British order being placed in August 1915.

Twenty-one firms, fourteen of them in France, were turned over to the manufacture of the revolutionary new 'eight'. Along with such concerns as Ariès, Brasier, and (ironically) Peugeot, Wolseley in England and Wright in America joined in the programme. One hundred thousand tons of raw materials, four-and-a-half million gallons of petrol, and 35,000 workers contributed to the manufacture of nearly 50,000 engines, representing 50 per cent of the total of all other types turned out on the Allied side during the war. Perhaps the best-known aircraft to use the Hispano were the British SE 5a and the French SPAD— 8,472 SPAD XIIIs alone were delivered. Output rose to 180bhp with the E-series unit which was quite happy at 18,000ft and could outperform its German rivals at this altitude. Birkigt also did some valuable work on interrupter gears which allowed fuselage-mounted machineguns to fire through the airscrew arc. Such devices eliminated unhappy improvisations like the SE 5a's secondary armament—a Lewis gun mounted on the top plane which was awkward to manipulate at any time, and almost beyond the capabilities of a pilot flying at high altitudes without oxygen.

The V-type Hispano was progressively developed in the inter-war period, four-figure outputs being achieved in the 1930s. The engine's successes were legion: the Gordon Bennett Cup in 1920, the Coupes Lamblin and Deutsch de la Meurthe in 1922, the world's altitude record in 1923, and the world's landplane speed record, which fell to Bonnet's Bernard monoplane in 1924 at a formidable 277·805mph, and stood for some eight years. Hispanos spanned the Atlantic on several occasions: Costes and Le Brix in 1927 and two crews (Jimenez/Iglesias and Assolant/ Lefèvre) in 1929. In 1931 a Hispano-powered Latécoère monoplane held the world's straight-line and closed-circuit distance records.

This distinguished war-time record in the air was commemorated on all post-1918 cars by the famous *cigogne volante*, emblem of Georges Guynemer's SPAD *escadrille*. It was Birkigt, a personal friend of the French ace, who commissioned the original silver-plated stork from the sculptor Bazin as a tribute to Guynemer's memory.

Though car production continued in Spain during the war years, no new models were announced, and perhaps the most significant feature was the standardisation of electric lighting and starting, which had been available as early as 1913—possibly a 'first' for Europe, though this is open to question. A new factory at Guadalajara handled the production of lorries and aircraft parts for Spain's colonial armies in Morocco: they also made the little 8–10hp L-head car. Oddly enough, the 'Suiza' part of the name was deleted from the radiator badge in this case only, giving rise to the rumour that Birkigt was not responsible for the model. In fact he was, but at the same time he was working on prototypes of his greatest car, the 32CV H6, or Type 41, as it was known in Spain. The genesis of the 32CV actually spurred the usually reticent Swiss into public speech at a banquet given in his honour by the Spanish Aero Club in February 1919.

He dismissed himself apologetically as 'only an engine designer', but what he said was of interest, in that it showed his thinking at the time. He made no bones about the new model's aeronautical inspiration, pointing out at the same time that the V-8 in its turn had been inspired by the 'España' of 1911. He further reminded his listeners that future cars would benefit from his aero-engine programme, while he paid a generous tribute to his supporters in Spain, notably the King and the Air Force.

That November the 32CV was unveiled at the Paris Salon. It was certainly the most significant design of the first post-Armistice decade. The unit gearbox, valve gear, and screwed-in steel liners were to be expected, but other departures were the dual coil ignition—Birkigt had been using dual ignition since 1907, but always previously with a magneto—and really efficient four-wheel brakes actuated by a gearbox-driven servo. These were far superior to any system then on the market, and were taken up by Rolls-Royce in 1924. As for the now traditional overhead camshaft, its influence was to be felt in an expected quarter, but to be developed in an unexpected way. Wolseley of Birmingham, who had made Hispano engines during the war years, applied Hispano principles to their inexpensive 1,267cc 'Ten' announced in 1919. This was hardly an outstanding vehicle, though it raced to some purpose, but when Wolseley succumbed to over-expansion in 1926, they were taken over by W. R. Morris, who was only too delighted to be presented

Page 153 (*above*) Marc Birkigt, a photograph taken in 1916; (*below*) Marc Birkigt in
later years

Page 154 (above) Racing *voiturette*: Zucarelli on his way to winning the 1910 Coupe de *l'Auto* at Boulogne on the four-cylinder Hispano-Suiza. The mechanic is uncomfortably close to the stub exhausts; (below) King Alfonso XIII of Spain at the wheel of an early 32CV torpedo with Spanish coachwork

Page 155 (above) The Town Carriage idiom: an Hispano-Suiza by Kellner, 1931; (centre) an Hispano-Suiza cabriolet by Fernandez et Darrin, 1931; (below) a 54CV twelve-cylinder Hispano-Suiza with bodywork by Saoutchik

Page 156 (above) Birkigt's last masterpiece: a 1934 Type 68-bis with 11,310cc twelve cylinder engine, when owned by the late Peter Hampton; (below) The engine of Peter Hampton's Type 68-bis V-12 Hispano-Suiza, a classic example of under-bonnet sculpture.

with a ready-made ohc unit—especially as Wolseley's range by now included a promising 2 litre 'six'. Not that the Nuffield Group's earliest ohc engines were an unqualified success—nobody loved the original Morris 'Minor' of 1929 and the Wolseley Straight-Eight was an infelicitous exercise—but the former served as the basis for the classic MGs of the Kimber era, while Wolseley's long connection with the police force was forged by the 'Fourteen' of 1935, the last of the original overhead-camshaft line. By a stroke of irony, too, the MG wrested domination in the small sports-car class from a French design, Emile Petit's twin-cam Salmson first seen in 1921.

To the world at large the 32CV and its bigger stablemate, the 46CV 8 litre of 1923, spelt Hispano-Suiza and thus Marc Birkigt. For all the limitations of a three-speed gearbox and a multi-plate clutch that was not up to the rest of the car, they looked, went, and stopped better than anything in their class, and even in 1925 speeds of 80mph were normally the preserve of specialist sporting machinery. There were few major alterations before 1928, and the cars were still available in France as late as 1934, the 46CV surviving two years longer in Barcelona's catalogues. But though Birkigt had progressed from the sporting to the overtly luxurious, he was still versatile, and something simpler was wanted for Spain, even if France might reject the lesser breeds. No four-cylinder cars were sold north of the border after 1914, and even the excellent 3·7 litre six-cylinder 'Barcelona' (which would see 100mph in later, three-carburetter form) was dismissed airily as *un loup* after a solitary Salon appearance in 1924. Its four-cylinder stablemate, the Type 48, was, however, a classic illustration of how Birkigt could apply his uncompromising standards to a humbler idiom.

Such a car could hardly have happened elsewhere. Spain, with no middle class as in other European countries, could not support a 2½ litre family tourer of the type prevalent in Britain, and such a machine would hardly commend itself to the aristocracy. Thus the only possible customer was the Government, and the Government's demands were stringent. A design competition organised in 1923 was aimed at evolving a vehicle for official use by the Army, the Department of Public Works, the Justice Department, and provincial governors, and these personages required a car that would give trouble-free service for more than a million kilometres. They got it.

The Type 48 was really only a four-cylinder version of the big ohc 'six'. An output of 60bhp from two-and-a-half litres was unremarkable, and the '48' was as remote from the general public as any Russian ZIS or ZIL of a later era. Occasionally the model would turn up at a foreign show—towards the end of the Vintage era it was seen both at Olympia

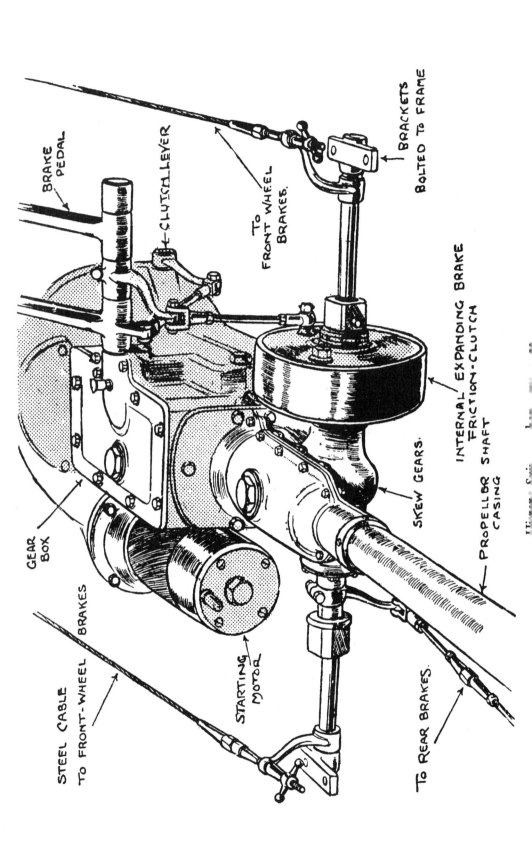

BRAKE PEDAL

CLUTCH LEVER

TO FRONT WHEEL BRAKES.

BRACKETS BOLTED TO FRAME

GEAR BOX

STEEL CABLE TO FRONT-WHEEL BRAKES

STARTING MOTOR

SKEW GEARS.

INTERNAL EXPANDING BRAKE FRICTION-CLUTCH

PROPELLER SHAFT CASING

To REAR BRAKES.

and at the Grand Palais—but it plodded on in government service through revolution and civil war. Production continued until 1935: the Governor of Valencia kept his from 1929 until 1941.

By now, of course, the Hispano-Suiza was seen more frequently at *concours d'elegance* than on the circuits, though there were still some impressive wins, such as those of Dubonnet in the 1921 Boillot Cup at Boulogne, and Bablot at Monza in 1922, on the high-performance short-chassis 6·9 litre 'six' which took its name from the Italian track. Output of the engine went up from 135bhp to 150bhp, which gave the 'Monza' two-seaters a top speed of 107mph on a 2·8:1 back axle, allegedly at the price of some rather tricky handling. The year 1923 saw the advent of the 8 litre with Garnier's class win at Boulogne, while in 1924 Dubonnet actually managed sixth place in the Targa Florio, at the wheel of 35cwt of tulip-wood tourer which consumed tyres at an alarming rate. But perhaps the most impressive performance of all was Weymann's victory in the celebrated match against a straight-eight Stutz at Indianapolis in 1928. The American car was appreciably smaller and less powerful, retailed for one-fifth of the Hispano's price, and was unlucky to be dogged with valve trouble, but against this the French aviator's 8 litre was a nine-year-old design, as well as being a well-worn example that had received no special preparation.

These great cars sold well, but Birkigt recognised that there was only a limited future for behemoths in the £2,000 price-bracket. He therefore initiated negotiations with Hudson Motors of Detroit, then riding the crest of the wave as the strongest of America's independent manufacturers. In 1929 they delivered 300,962 cars, and commanded a 6·6 per cent share of the home market. Further, their 'Super Six' had an excellent reputation for toughness, and was scheduled for early replacement. For Hudson, Birkigt designed a simple short-stroke pushrod 'six', the 3 litre Type 60, with a single-plate clutch which marked a great improvement on previous models, and Lockheed hydraulic brakes incorporating the famous servo. This was a modest car on a 10ft 8in wheelbase, and might have gone down well at a time when American luxury models were reaching elephantine proportions. Alas! The economic situation nipped this project in the bud, though the '60' was made in Spain from 1932 until the end in 1943.

In 1928 the big 'sixes' were brought into line with current aero-engine practice. The principal improvement was the use of nitralloy cylinder liners: the more effective hardening achieved by this process allowed engines to operate with impunity at far higher temperatures, though both time and cost of manufacture were increased thereby. Such refinements were, however, entirely viable despite the stock-market crash, for universal disarmament was by now a lost cause, and the demand for aero-engines remained steady.

Bois-Colombes might be fully employed, but the situation in Spain had lapsed into chaos. The revolution of 1931 cost King Alfonso his throne, and Birkigt one of his staunchest champions—they had been firm friends since the designer was presented to the monarch at the 1907 Madrid Show. This process of disruption was completed by the civil war of 1936. The factories at Barcelona were severely damaged, while

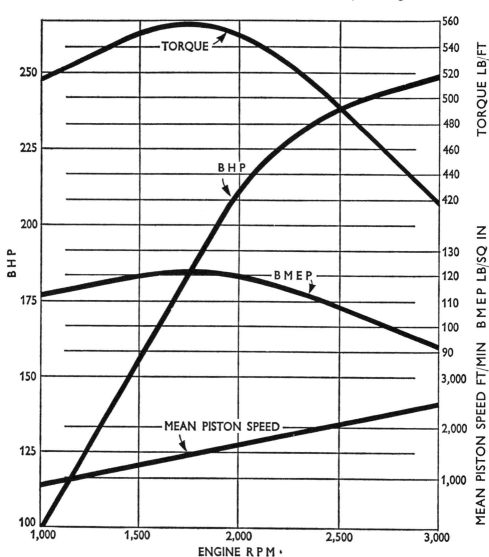

Power curves for the 11·3 litre Hispano-Suiza

Guadelajara virtually disappeared, since the Republicans transferred all the machinery to Valencia, and much of it was irrevocably lost. There was, however, some compensation when a new Hispano-Aviacion works was opened at Seville, in Nationalist territory, in 1937.

Back in France, the magnificent twelve-cylinder pushrod Type 68 unveiled at the 1931 Salon showed that Birkigt had not lost one iota of his touch. This herculean 9·4 litre carriage was capable of over 100mph, even if such performance involved a bottom-gear ratio higher than the top gears of some small family saloons, which could render re-starts on steep hills a trifle sticky. The pushrod engines were quieter than the ohc types they superseded, and the modern 100 × 100mm engine developed 220bhp, later raised to 250bhp when an 11,310cc unit was made optional in 1934. Unfortunately the V-12 was the sort of car that only an aero-engine factory with a full order-book could afford, for the English price with a fairly simple body was £3,750, and the annual tax was £75. Admittedly, a short-chassis coupé weighed less than two tons, but the carriage-trade was looking for something a little more compact, and even Birkigt's last new model to see series production, the 30CV K6 of 1934, was still a lot of motorcar. It was perhaps fortunate for Birkigt that Adolf Hitler's sabre-rattling focused attention on the deficiencies of France's air force.

By 1938 the 'regular' 12Y Hispano-Suiza aero-engine was rated at 930hp, while another interesting development was the *moteur-canon*, which allowed for the installation of large-calibre guns on fighter aircraft without the complication of wing mountings. On *canon* versions of the Hispano engine a 20mm cannon, also of Hispano design and manufacture, fired through the airscrew boss, and this unit was fitted to two of France's standard fighters, the Morane-Saulnier Ms 406 and the Dewoitine 520.

Unfortunately, French industry was thoroughly run-down by the 1930s and the country never recovered from the Occupation of the Factories in 1936. Nothing that Birkigt could do succeeded in expediting deliveries of sorely-needed engines, even after the discontinuation of the cars in 1938. There were always airframes awaiting power units, and in the immediate pre-war period French aircraft makers were feverishly shopping for Hispano-type aero-motors in Switzerland and the USSR. They even tried to do a deal with Nazi Germany for the supply of Daimler-Benz inverted-Vs. Birkigt tried hard: there were new research laboratories, new factories for the production of guns and ammunition, the world's most modern foundry at Bois-Colombes, and a new wind-tunnel. A factory at Tarbes supplemented the complex in Paris, and during the last two pre-war years only military contracts were handled.

But though no new Hispano-Suiza cars were made, there was a dis-
creet approach from Rolls-Royce, as the Marquis de Salamanca recalls.
Salamanca, the Rolls-Royce concessionaire in Madrid, had spent many
fruitless years endeavouring to convert Alfonso XIII to 'the Best Car in
the World'. The King actually tried a 'Silver Ghost' in 1910, but he was
unconvinced. Even Salamanca's victory in the Spanish Touring-Car
Grand Prix of 1913 (which Hispano-Suiza did not support) failed to
shake the royal allegiance, despite many a verbal battle between subject
and sovereign.

According to Salamanca, the powers that be at Derby approached
Birkigt in 1934 in the hope that he would assume the mantle of Sir
Henry Royce: in view of the imminence of a twelve-cylinder car, the
choice was a natural one. The Spanish nobleman actually offered to
organise a meeting at the Hotel Scribe in Paris, but this never took
place. Salamanca believes that the opposition came, not from Birkigt,
but from the French government, who put pressure on him to stay away.
One must confess to a private feeling of regret, for a Hispano-Suiza
engine allied to the peerless Rolls-Royce four-speed synchromesh
gearbox would have resulted in a car to end all cars—provided, of
course, that anyone could have afforded it!

Nor was this the first occasion on which Salamanca had served as a
go-between, for in 1923 he installed Hispano-Suiza servo brakes on a
'Silver Ghost' belonging to the Countess of Prado-Ameno, and took it
to Le Canadel to show to Royce. The English *maestro* was both sceptical
and annoyed at the sight of foreign appendages on one of his cars, but
within a year he had adopted the system. Salamanca also remembers a
meeting between Royce and Birkigt at the 1919 Salon. The 32CV was,
of course, the star of that show, so one need not be surprised to hear
that Royce told his Spanish agent that he would 'like to have had
M. Birkigt's imagination'.

The outbreak of hostilities in September 1939 saw some frantic and
belated expansion in France. Birkigt opened a new underground factory
at Janzac to try and keep the production lines flowing. When Paris fell,
he transferred his headquarters to Tarbes. After the Armistice, of course,
all work came to a standstill, but meanwhile the 20mm Hispano cannon
had gone into production in Britain, where it was fitted to Spitfires,
Hurricanes and Beaufighters. More than 90,000 of these weapons were
delivered, along with one hundred million rounds of ammunition.

Birkigt flatly refused to collaborate either with the Nazis or with the
Vichy Government, but as a neutral citizen he was free to return to
Barcelona, where he devoted himself to the rehabilitation of a Spanish
Air Force weakened by nearly three years of civil war. A fuel-injection

version of the 12Y engine was persuaded to give 1,300bhp, and in 1944 he sought to ease another wartime shortage by designing a modern forward-control diesel lorry. This came through its trials with flying colours, but problems with raw materials prevented series production. After the war, of course, Spain was subjected to an economic 'blockade' which further impeded her recovery, while the sale of the Barcelona factories in 1946 brought in a new team under Wilfredo Ricart. Ricart was also a commercial-vehicle specialist (he had worked on diesels during his years with Alfa Romeo), and he discarded Birkigt's theme in favour of one of his own creation, the first of the Pegaso line.

Once again Birkigt went back to Bois-Colombes. But times had changed, and the situation across the border was even more parlous. The factories had been heavily bombed by the Allies in 1943, and the staff had dispersed. Further, Hispano-Suiza derived no benefit from Marshall Aid, the fairy godmother that was to save Alfa Romeo, among others. Not that anyone claimed that Hispano had 'collaborated' in the manner of Renault or Berliet. Nor had they failed to go slow enough on German orders, a charge levelled against Hotchkiss. The trouble was simpler: a large number of the shareholders were Spaniards, and Spain's attitude to the Axis Powers had scarcely been ambivalent. French and Spanish observers alike gained the impression that a healthy and prosperous Bois-Colombes might have furnished a little too much competition for American industry at a time when the aero-engine factories at East Hartford and Paterson were having contracts cancelled right and left. In a country whose motor industry is the world's biggest closed shop, 'going into cars' offered no escape route, as Henry Kaiser was to discover to his cost. Not that Birkigt was forgotten—the Aero Club of France awarded him their *Grande Medaille d'Or*, and he was elected a Grand Officer of the Legion of Honour. In his native Switzerland, Zurich's Ecole Polytechnique Fédérale made him a doctor *honoris causa* in 1945.

If the American aircraft industry could not save its bacon with cars, Hispano-Suiza tried once more, though their principal effort was a medium-priced 4 litre ohv V-8 with front-wheel-drive of which Birkigt wanted no part. Only one prototype was built, and this did its road trials with a 3,622cc Ford V-8 engine under the bonnet. Birkigt's own ideas ran to something smaller and more complex described as another *merveille d'horlogerie*, but this one was equally ill-suited to the uneasy climate of early post-war France, and he retired to his estate at Versoix, outside Geneva. Here he died in March 1953. Only a few old friends attended the funeral.

To the end, he remained an unknown and the motoring public only became aware that he was a family man when one of the few 'private' Type 48s was supplied to his daughter in 1929. However, he seems to have been more sociable than many dedicated folk, and enjoyed his family circle. He had no hobbies, though he sometimes sought recreation in hunting, walking and the music of Wagner. Spain appealed to him for its culture rather than its social life.

Within his factories he was a perfectionist. The cheap and the shoddy he rejected utterly, and cost accountants were anathema to him. He regarded the American concept of planned obsolesence as immoral: hence his cars tended to be ahead of their time and ageless. His 1904 theme (which inspired the 'Alfonso') carried him up to World War I, and for the next fifteen years he occupied himself with developments of his 1911 overhead-camshaft engine to suit the carriage trade of France and Spanish officialdom. The car he unveiled in 1919 was still largely acceptable in 1935, and it was the immense cost and size of the pushrod V-12 rather than any inherent archaism of design that killed it. Unfortunately, any replacement designed to Birkigt's uncompromising standards would have been either too small for his tastes, or too expensive for a market weakened by the Depression, and he could never have worked in an idiom that was foreign to him.

Genius has been defined as 'an infinite capacity for taking pains', and if we accept this definition Birkigt certainly qualifies. One hesitates to name any designer as the first man to do anything, but we may well owe the unit construction of engine and gearbox to Birkigt, and certainly he pioneered the gearbox-driven servo brake. The solid Birkigt may have had the appearance of the Genevese Calvinist that he was. He was tireless: but tirelessness can exhaust and ultimately exasperate others, and Birkigt was no hair-tearer. He could contemplate a problem in silence for hours on end, and he would follow each project to its logical end. He regarded both the method of manufacture and the necessary machines as part of a design job. In the case of the aero-engine, this led logically to the interrupter gear, and thence to machine tools capable of producing this.

The quiet genius is hard to resist, as is the man who knows what he wants, and one of the most intriguing aspects of the Birkigt story is that this self-effacing man possessed a gift for getting on with people. Mateu, Massuger, Barron, and Alfonso XIII himself all liked and respected him. Birkigt even went hunting with the king. But perhaps his versatility is his most interesting facet, for it is hard to think of another car designer of the front rank who was also responsible for one of the outstanding aero-engines of one war, and the most significant air-to-air weapon of another.

6 FERDINAND PORSCHE

by D. B. TUBBS

FERDINAND PORSCHE (1875–1952). Austrian:
Sometime designer for Steyr, Austro-Daimler,
Daimler-Benz (Mercedes S-type), Auto Union
(GP racing car), Volkswagen, Cisitalia (GP racing
car) and Porsche, as well as successful aero-engines
in both world wars. Probably the most versatile
automobile engineer yet.

FERDINAND PORSCHE

IT HAS BEEN said that most designers are responsible for only one, or perhaps two, really outstanding designs. Porsche clearly disproves this dictum, overtopping the rest of the engineering world like one of his own overhead camshafts. His genius was universal. He busied himself with and mastered every branch of automobile design during a creative lifetime of more than forty years, so that even to epitomise his output would overburden this opening paragraph. Let us first salute him therefore as the creator of just half a dozen brilliant machines: the Prince Henry Austro-Daimler, 38/250 Mercedes-Benz, sixteen-cylinder Grand Prix Auto Union, and Volkswagen among a host of automobile designs, together with the Tiger Tank in World War II and the most successful aero-engine used by Germany in both world wars. We may also remark that Dr Porsche was the best rally driver of his day, and a redoubtable, hard-swearing character who was apt to say what he thought. He also patented torsion-bar suspension, and was a pioneer of petrol-electric transmissions, for, like F. H. Royce, he was an electrical engineer before he turned to motorcars.

Porsche was a Bohemian; that is to say he was born at Maffersdorf on the northern frontier of Bohemia on 3 September 1875, when that country was a separate kingdom forming part of the Austro-Hungarian Empire. Later, this bit of Europe (known to recent history as the Sudetenland) became part of Czechoslovakia, and Porsche, fervently loyal to his home town, opted for Czech nationality. This was fortunate, because it later enabled him to motor about Europe testing cars and visiting foreign exhibitions when German citizens found their right to travel restricted. Physically, as well as mentally, Porsche liked to keep on the move.

The name Porsche is not German, but a corruption of the Slav name Borislav. Ferdinand was the third child of a fairly prosperous master tinsmith owning his own business; rather a no-nonsense character, one gathers. This Plain Man objected to young Ferdinand's wasting his time on electricity after a twelve-hour day at tin-smithing, and there were particularly ugly scenes one day when he kicked over one of the boy's accumulators, not knowing, poor mensch, that batteries have acid inside. Although Papa Porsche burned his boots, Mama took her son's part. Obviously the boy had talent, so why not send him to technical college? If Vienna was too far and too expensive, let it be night-school at Reichenberg, the nearest big town.

The fifteen-year-old went joyfully off, and it was not very long before

Papa came home one night from a business trip to find a great radiance shining from his house; Ferdinand had installed the electric light as a surprise, having made everything—generator, switchboard, wiring and all, with his own hands. The only other electric light in Maffersdorf at this time was at the Ginzkey carpet factory, where Porsche's brother-in-law worked. An interview was arranged, and the Ginzkeys, impressed, sent young Ferdinand off to Vienna with their blessing and an introduction to the great electrical firm of Bela-Egger, which is now Brown-Boveri. This meant not only a job, but a chance to attend courses at the Vienna Technical College. Joyfully he profited by both; promotion was rapid and inside four years young Ferdinand found himself head of the test-shop and chief assistant in the calculating department. It was time for a move.

Although the world's first motorcar had been made by a Viennese in 1875, the year of Porsche's birth, no Austrian followed Markus's example until the late 1890s, after Daimler, Benz and Panhard had shown the way. Even the 'rattling, stinking mosters' were deemed out of place. Something was needed that could show its face at Court, and waft pretty ladies home after the opera; so at least thought Lohner, head of an old-established firm of coachbuilders to the Austro-Hungarian court. Hearing of young Porsche, he offered him a job and the task of designing a silent electric carriage.

This was very much to the young man's taste. 'The trouble with most horseless carriages,' said he, 'is the complexity of their transmission. All this business of shafting, bevel gears and chains could be quite simply avoided if we used electric cables to carry the power to the place where it is needed, which, obviously, is at the driving wheels.' Visitors to the Paris Salon of 1900 were able to see what he meant. There, on the only Austrian exhibit, they could study the System Lohner-Porsche in an electric chaise of the young man's design. It was driven by storage batteries and there was an electric motor in each of the front wheel hubs. Press and public were enthusiastic, especially when, back in Austria, young Porsche started setting hill-climb records with a stripped racing version of the car. On 23 September 1900, for example, he broke the record for the Semmering Pass with an average of 25mph for six miles of very twisty uphill road. This was easily the best performance recorded by an electric vehicle, and the car is interesting as being the first successful motorcar with front-wheel drive, and the first to have four-wheel brakes, since the motors in the front wheels acted as electrostatic brakes on the over-run and there were the usual drum brakes on the back wheels. Mr Porsche's skill as a driver was widely applauded as well.

The trademark of the System Lohner-Porsche

Both Porsche and Lohner quickly realised the two disadvantages of an all-electric car: heavy batteries and limited range. Their answer was the 'Mixte' transmission, soon adopted and made under licence by Mercedes, in which a petrol engine was used to generate electric power for transmission by cable as before to motors in the driving wheels. There were clear advantages in this system at a time when ham-footed coachmen were battling with an ill-maintained scroll or cone clutch, and when even the best gearbox and side chains made a good deal of noise. So marked was the smoothness indeed, that a Mercedes-Mixte was chosen as his staff car by the Archduke Franz-Ferdinand in the military manoeuvres of 1902, and the Archduke (later to be murdered at Serajevo in 1914) was so pleased that he sent a letter of congratulation to its chauffeur on that occasion, a certain army reservist named Ferdinand Porsche, stating that 'the performance of your car together with the safety and precision of your handling of it were much appreciated by his Imperial and Royal Highness'.

Porsche took out his first patent on the Mixte system while he was still with Bela Egger in 1897, and although it is tempting to dismiss it as one of the primitive 'funnies' of the first decade, the system played a sometimes decisive part in two great wars, and gave rise to some of the most spectacular and unlikely vehicles ever built. These are worth looking at before we return to motorcars proper and Porsche's prowess as designer and competition driver of the Edwardian period. During that time Porsche was busy on work for the army, and in 1910, when he might well have been too busy with the Prince Henry Trials, he designed his fantastic Land-Train based on the Mixte system. The train comprised a 'locomotive' tractor and anything up to ten trailers. The locomotive was fitted with a 100hp petrol engine driving all its own wheels by means of electric motors in the hubs, while current could also be fed to hub-motors in each of the trailers. It may be asked how such a long train managed to get round corners, but this Porsche had also foreseen, having invented a form of self-steering tow-bar. The cumbrous equipage could thus cope with quite sharp bends, and was in fact driven over mountain passes in the Italian Alps, and could even be steered in reverse. If very sticky going was encountered or the train had to cross bridges that would not take the weight of more than one wagon, the rest of the train was uncoupled and the tractor stood on one side feeding current through long cables to the hub-motors of each wagon in turn until the whole lot had crossed. This land train scored a tactical success by bringing supplies through apparently impossible country during the Isonzo campaign on the Italian front in 1917.

The C-Zug was a similar machine, but its task was the transporting of a single enormous siege-mortar. This monster was nothing less than a 420mm (16·5in) gun, which still remains the heaviest gun ever to travel by road. The barrel alone weighed 16 tons, 26 tons with the rest of the gun. The limber weighed just as much, and there was the cannon-bed that weighed still more. One round for this enormous gun weighed 20cwt. The load was spread between six self-contained units, each comprising a 150hp generator tug and one eight-wheeled wagon, and as both back wheels of the tug and all eight of the trailer had hub motors each unit possessed ten-wheel drive. The all-up weight of each unit was 53 tons. Porsche never forgot the success of his Mixte system, and he was to use it again during World War II in the design of his Tiger tank. Lovers of the picturesque will always regret that his 1917 plan for a captive aircraft on the same lines—something between a helicopter and an observation balloon fed with current from a generator on the ground —was shelved at the end of the war. This would have been quite something to watch.

It must be confessed that the stream of novelties bubbling from Porsche's brain was always something of an embarrassment to his employers. The counting-house never loves a perfectionist, and the Doctor was to find this all his life. Fortunately, by the time that he set up on his own there were numerous clients to occupy his mind and a steady income from torsion-bar royalties to keep wolves from the door. This, however, is looking many years ahead; the technical agility of Porsche in his twenties was too much for Lohner, the Viennese carriage-builders, and after seven stimulating years with them, during which he was awarded the Pötting prize for engineering studies, he was invited to join Austro-Daimler in the autumn of 1905, where he was at once appointed managing director, in succession to Paul Daimler who had held that post since 1901.

Austro-Daimler's main line in the years 1905–8 was the Maja, a 30hp car named after the second Miss Jellinek, whose sister Mercédès had already been immortalised by the German Daimler company, and before designing a replacement for it Porsche went to work on a racing-car for the Kaiserpreis, using a Mercedes engine and the Mixte transmission; it was to be made in Stuttgart. Dual ignition, 85bhp, 87mph and water-cooled brakes. The Press were much impressed by this machine and also by the new, light, clean and airy factory buildings that Porsche had put up. Reporters found him, to their glee, in the racing department, screwdriver in hand '*pour bien régler le moteur*' in one of the hubs, but he might equally well have been at his drawing-board at work on an aero-engine, or, indeed, up in an airship, for already—two years before Blériot flew the Channel—Porsche was very busy on engines for air-craft. And, incidentally, the technical knowledge and cool head of the young engineer came in very useful one day when the gas-release valve jammed in a Parsefal 'blimp' in which he was OC engines. Porsche climbed up the rigging and freed the valve before the envelope could burst. A near thing.

The Maja's replacement, a 32hp with four-speed box and choice of live axle or chain drive, was a success commercially, and earned for Porsche his first decoration, an Order from the King of Bulgaria. A team was entered for the first Prince Henry Tour in 1909. All cars came through without penalty marks and performed well in the special tests, earning a silver plaque. Porsche himself drove one car, but vanished before the prize-giving. A co-director ran him to earth in a writing-room at the hotel. 'What on earth are you doing here?', he cried. 'Designing a *proper* car for next year', said Porsche, and this was no overstatement, for in 1910 Austro-Daimlers finished first, second and third in the Prince Henry with Porsche and his wife in the winning car, won nearly all the special tests and showed themselves a clear 10mph faster than their nearest rival, with a top speed of 87mph from 86bhp. The long-stroke 5·7 litre engine was a very advanced design, with inclined valves operated by a shaft-driven overhead camshaft, and the body form was strikingly original, the sharply tapering sides of its 'tulip' body giving an almost irreducible cross-sectional area for a four-seater car. The Prince Henry Austro-Daimler had tremendous success in competitions, and the engine layout was the basis of Porsche's many aero-engines used before and during the 1914–18 war, after which conflict a British aeronautical paper described the A-D as quite the best aero-engine to see service with the Central Powers.

Here it must also be confessed that the ohc Prince Henry car and its production derivative with live axle, sold in Britain as the 27/80, were exceptions in the A-D range, which included some extremely dull side-valves with dismally 'alpine' gear-ratios. Some of these aped their betters, having sharply-pointed radiators and tulip-shaped bodywork.

Shortly before 1914 the A-D company merged with the great arms firm of Skoda, and Porsche was more busy than ever on military designs. Some of these, like the Gun-tug, have already been discussed; another, known as the Daimler Horse, may be mentioned here as it introduces one Karl Rabe, who was to become Dr Porsche's right-hand man. The 'horse' was a two-wheeled tractor designed to hitch on to such things as light cannon or field-kitchens, and the immediate problem was to evolve all-metal wheels which would give a good grip in the mud of the Eastern Front and yet not wear out quickly on hard roads. Retractable cleats were the answer, but how to make them retract? The whole drawing-office wrestled with the problem, until finally Porsche offered a personal reward of 500 krone to anyone finding a usable solution by the following day. Karl Rabe, a newcomer of only twenty, showed his idea to the foreman, but the latter was unimpressed. It looked too simple; however, he did agree to pass word to the boss. Porsche stood behind the young man's drawing-board for what seemed an hour. Then 'Pay out! Pay out!' said he, and left the room without another word.

This was typical of the man. He was always on the prowl, looking over shoulders in the drawing-office, so that rude people used to say that he couldn't draw a line himself. His method was to argue things out with his assistants. 'That's a load of muck' was his stock response, and out of the ensuing argument a solution would emerge. His methods, and his language, were apt to startle strangers. When he first went to Austro-Daimlers, people were not used to a boss who wandered about the shops, Paul Daimler having been rather Olympian and inaccessible. One day a car had been brought into the experimental department with some obscure trouble. Engineers in clean white coats were clustered around in discussion when Porsche happened along. Quietly he borrowed some oily overalls and crawled underneath. Great confusion when he was recognised on the way out again. 'Ach, Herr Direktor,' cried a white coat, 'now you can tell us what is wrong.' 'Find out for yourself you —— —— ——!' said the Herr Direktor, using words the engineer had never learned at home, and was not seen for the rest of the day.

Page 173 (right) Dr hc Ferdinand Porsche (1875–1952); *(below)* in the Porsche 'mixte' system, a petrol engine was used to generate current which was fed to electric motors mounted in the hubs

Page 174 (above) One of Dr Porsche's first great successes, the big Austro-Daimler with ohc engine, built for the 1910 Prince Henry of Prussia Trials. He is here seen at the wheel of the winning car, with Frau Porsche among the passengers; *(below)* as a post-war venture at Austro-Daimlers, Dr Porsche built this successful twin-ohc, 1,100cc four-cylinder sports and racing car. The model was named 'Sascha' after the well-known Austrian sportsman and film producer, Count Sascha Kolowrat

Page 175 (above) The Mercedes-Benz 38-250 SSK, here photographed at Shelsley Walsh hill-climb where Rudolf Caracciola set a new sports car record in 1930; *(below)* Hermann Mueller at the wheel of the Auto Union GP car, Type C, during the 1937 Coppa Acerbo at Pescara. Note the Porsche trailing-arm ifs. By 1937 the supercharged V-16 pushrod engine had increased in size to 6,330cc and was giving 545bhp

Page 176 (*above*) The rear-engined 1·4 litre Porsche 32; (*right*) the Cisitalia GP car designed in 1946; (*below*) the first model to bear Dr Porsche's own name, developed in Austria in 1948

The Great War brought Porsche a number of official honours, but the one which pleased him especially was the honorary doctorate conferred upon him by the Vienna Technical College. It made him feel that the night-school boy from Maffersdorf had 'arrived'. Certainly there was plenty to do—a whole post-war programme to prepare in the very difficult circumstances of post-war Austria, now reduced from an empire to little more than the capital city and a rural hinterland. The problem was one of exporting, as in Britain after the second war. With Rabe in charge of the drawing-office, promoted over the heads of older men, Porsche produced a big 4·4 litre six, giving 60hp. The overhead-camshaft engine was strikingly architectural in appearance; Porsche always maintained that the outside of engines was important. For the home market there was clearly a demand for small cars, and Porsche, true to form, decided to begin with a sports car and work down. He was fortunate in finding a friend and patron in one Count Kolowrat, a wealthy sportsman who mixed motor racing with film-producing and similar glamorous pursuits. Sascha Kolowrat's name was attached to the chosen design, and the little cars were wonderfully successful, for the overhead-camshaft engine of only 1,100cc gave considerable power, and the two-seater Sascha (which was a 'sports car' when running with wings and a racer without, like Bugattis of the same period) was timed over a kilometre in Holland at very nearly 90mph in 1922. Three Saschas were entered for the Targa Florio, two in the 1,100cc class and one whose bored-out engine placed it among the big racing cars. This was to be driven by the chief tester at Austro-Daimler, one Alfred Neubauer. In the race the eleven-hundreds won their class against strong French and Italian opposition, and Neubauer finished honourably high in the general classification, with only five cars ahead of him. The Sascha had a considerable *succés d'éstime* in the international motoring papers, and at the same time Porsche was busy on a range of larger racing cars, including a 2 litre for the 1924–5 GP formula. Having watched the impact of his Prince Henry victory before the war, Porsche was convinced that the way to the buying public's heart lay through racing. Certainly he did well with the Sascha; in 1922 the Works had forty-three wins and eight seconds from a total of fifty-one starts.

There was at this time a strong anti-racing faction on the A-D board who thought that Porsche was scattering his fire. They pointed to a bewildering array of changing models, and were apt to tell a story about a customer who brought his six-months'-old car to the service station, only to be told that no spares were available for such an obsolete model. There was something in this, and when the Works driver Fritz Kuhn

was killed at Monza when a wheel collapsed on the 2 litre, the accident was made a *casus belli*. This was in the spring of 1923.

Porsche was out, but he was immediately invited to join Daimler-Benz AG, the Mercedes company, with a seat on the board. He did not find things easy once he was there. There is very strong local patriotism in Germany amounting to national pride, and the Austrian was anything but *persona grata*. Once more Porsche was stepping into Paul Daimler's shoes, and the Mercedes factory, like Austro-Daimler in 1906, noticed a great difference. They knew Porsche by reputation of course, but they found his informal, ubiquitous and hard-swearing presence a little too unconventional. They thought him no academic; in fact the Stuttgart Technical College refused to recognise his honorary degree from Vienna, saying that 'foreign' degrees could not be accepted. They went even further, to the length of requesting him to remove the honorary doctorate from his newly-printed visiting cards. It may have been some consolation to Porsche during this difficult time to see that Austro-Daimler managed to live for two years on the designs that he had left behind him, notably the 2·6 litre ADM with ohc, two carburettors, four-wheel brakes, and one-shot lubrication, which in 1924–5 notched up more than sixty wins in speed trials and rallies. This was later developed by Rabe (who remained at A-D for some years but joined Porsche when he opened his own shop and became right-hand man of Porsche's son when the Porsche car was designed) into the ADM-R as a hill-climb car that was to give Hans Stuck innumerable wins, and make him European Hill-climb Champion for 1930 in the racing class, ahead of a 2·3 Bugatti.

Meanwhile, at Mercedes, Porsche himself was busy with racing. The first-string was a four-cylinder 2 litre which had already raced at Indianapolis where, as Lautenschlager remarked, 'the prize-money got stuck in the *Auspuff*.' This may have been sour grapes, seeing that the speaker ran into a wall and another member of the team finished third, but the criticism was a just one: the manifolding did not suit the experimental supercharger, and it was Porsche's first task to make it do so. He worked for nine months, and the result was a brilliant and resounding victory for Mercedes in the next year's Targa Florio, where Porsche himself acted as racing manager. An international win was exactly what Germany wanted to re-establish her *amour propre*. Christian Werner received a civic welcome upon the team's return to Stuttgart; Porsche, too, was required to sign his name in the city's *Livre d'Or*, and what is more the Stuttgart Technical College acclaimed the designer by not only recognising his Viennese doctorate, but also conferring one of their own. The Swabians had at last made the Doctor feel at home.

There was much else to be done at Untertürckheim. From the Targa cars, Porsche developed a straight-eight with permanently engaged blower that brought the young Caracciola his first victory in the German GP, but the chassis was by no means good, and it was his 'touring' cars that brought Dr Porsche fame during the 1920s, by their ability not only to tour (as witness Caracciola's winning the Tourist Trophy), but also to win Grand Prix races. The improbable ancestor of this line was the 28/95, whose chassis was purest Edwardian, having appeared before the Great War, but whose 7·3 litre six-cylinder engine startled the Berlin Motor Show so early as 1922 by wearing a plug-in Roots blower.

From this design of Paul Daimler's, Porsche first discarded the engine, which had steel cylinders like the war-time aero-engines, and replaced it with a much more modern affair using separate liners in a light-alloy crankcase-cum-block, with detachable cast-iron head. The overhead camshaft lived beneath an aluminium cover and was elegantly driven by a concealed shaft and silent spur gears in the back of the block. The front of the camshaft drove an imposing cast-alloy fan lying directly behind a high and stately pointed radiator. Unlike the 28/95, which profited by its ohc to have inclined valves, these engines, which came in one design but two sizes (4 litre and 6 litre) had vertical valves. They were first installed in long touring chassis, but Porsche had other plans, so that in 1925 the short-chassis 'K' model ('K' for *Kurz*—short chassis) took thirty-seven speed awards. The series was launched, and the 'K' went into limited production for 1927 together with a more powerful model with lower centre of gravity called the 'S' (for Sport). These were the models sold in Britain as the 33/180 and 36/220 respectively, the latter having an underslung chassis and two carburettors. Its success was enormous, and in 1928 this 6·8 litre monster was joined by the 38/250 of 7·1 litres (the 'SS' or Super Sports) and a short-chassis 'SSK' claiming 225bhp with the blower engaged. The last word, used mainly by Works drivers, was the 'SSKL' ('L' for *Leicht*—light) with very much lightened chassis.

In sports car events, these machines won such races as the Tourist Trophy, the Irish GP and the Mille Miglia, while they also wiped the eye of full-dress Bugatti Grand Prix cars (Chiron up) in the *formule libre* German GP on the Nürburgring. They showed great reliability, and were extraordinarily light on tyres. Naturally, too, with their great size and splendid orchestral accompaniment of *Sturm* (from the tailpipe) and *Drang* (from the blower), they were most popular with the crowds. One likes to think that Dr Porsche chuckled sometimes to think that he, a 'foreigner' from Vienna, should have designed the most Teutonic conveyance ever conceived. He must also have been pleased in 1930, the year Rudolf Caracciola won the sports-car class of the European Hill-climb Championship, that another model which was basically his—and a sports car at that—should have won the *racing* class in the Championship for Hans Stuck. This was the ADM-R Austro-Daimler, which had been quietly but brilliantly developed in the meanwhile by his ex-assistant, Karl Rabe.

In discussing 1930, however, we are getting ahead of the story, to a point where Porsche and Rabe were to be reunited, and in a works of their own. In 1924, when Porsche began on the big 'touring' cars, German industry was just emerging from the disastrous inflation. Mercedes reached an understanding with Benz, her old rival, which was to lead to the merger of 1926, that was to make life so difficult for Ferdinand Porsche later on.

In sharp contrast to the overhead valves traditional at Mercedes, Benz were wedded to a side-valve policy which Porsche found very irksome. He found the 2 litre 'Stuttgart' and 3 litre 'Mannheim' models dauntingly 'worthy', from their flat radiators to their classic banjo rear-ends. The centre-change, three-speed box was as un-Porsche as it could be, and although as an engineer he may have applauded the seven-bearing crankshaft, an output of 19bhp per litre can hardly have been to his taste. Also, thinking already in Volkswagen terms, he wanted to launch a small cheap car for the masses—but neither board in the new combine saw the prole as a possible motorist. None the less he did get thirty 'prototype' 1 litre cars put through and a few of these were eventually sold.

It must not be thought, though, that Porsche became a 'frustitute' during those years with Mercedes-Benz. He was busy on Diesel engines —and won the RAC award in London for the best five-tonner of 1928. He also laid down the plans of an inverted 1,000hp petrol or diesel V-12 aero-engine which was to achieve fame in 1939–45 as the DB-600, collaborated with BMW in the design of the famous shaft-drive 500cc motorcycle, and embarked on very secret plans for a tank, which was built by Mercedes-Benz and tested at Kazan, in Russia, as Germany was technically forbidden to enter the arms race. For all this, tension within the Works grew, and it culminated in Porsche's being offered a penny to go and play in the next street. Daimler-Benz sent him on a fact-finding mission to the USA, with the offer of a consultant's job on his return. Instead of the latter, he accepted an invitation to join the Steyr group in January 1929, and moved out, letting his Stuttgart villa to Dr Nibel, who was to take his place.

To join this famous and old-established Austrian concern, with a tradition of craftsmanship in small-arms manufacture as well as some interesting automobile designs to its credit must have seemed an ideal solution. Steyr were especially interested in the springing problems raised by the bad roads of central Europe and already had a model running with independent rear suspension, the Type XXX. To this Porsche married an overhead-valve 2 litre engine that he had brought in his head from the unappreciative Mercedes. The resulting model continued in production for a number of years, but his masterpiece was the 'Austria', one of the most advanced machines of its day. Very much in the luxury class, it had a 5·3 litre pushrod straight-eight ohv engine with two plugs per cylinder giving 100bhp, a thermostat, independent rear springing by swing axles, and an overdrive gearbox in unit with the engine and using the same oiling system. Porsche again used cast-iron liners, and with its one-shot chassis lubrication the 'Austria' was one of the most up-to-date motorcars in Europe. Proudly Porsche drove off to the Paris Salon that autumn of 1929 in a handsome five-seater cabriolet with wire wheels; and within a few days, reading the paper in the lounge at his hotel, he learned that an Austrian bank had failed, and that as a result Steyr were to be taken over by Austro-Daimler. That meant the end of the 'Austria', and a change of address for Porsche. He resigned, and resolved to set up shop on his own.

The obvious place was Stuttgart. Here, in the middle of the motorcar country, he would be surrounded by specialist firms like Bosch, Mahler pistons, Hirth the crankshaft and engine people, together with a host of machine shops and foundries able to do any jobs Porsche might not wish to undertake in his own place at first. Financial backing came at this time from Adolf Rosenberger, an ex-racing driver friend who combined great commercial acumen with an enthusiastic love of the sport. He was quite a notable racing driver, mostly on Mercedes cars; in fact, he was right up in the Caracciola class in the late 1920s, finishing second to Rudi in the opening race on the Nürburgring in 1927, and beating that driver to a class win at the Freiburg hill-climb. He also won his class at the Klausen hill-climb and came second to Momberger in the unlimited class in the 1929 German GP, driving an SSK. This training in uphill work no doubt prepared him for his business partnership with Porsche.

It is difficult when probing into the latter's life not to think occasionally of Dr Strabismus, whom God preserve, of Utrecht, for he seems to have had a good deal in common with that zany but endearing savant of Beachcomber's imagination. Certainly Rosenberger did not find it easy to keep the Doctor's feet on the ground, and in the early days of the Porsche Büro at Kronenstrasse 14 it was sometimes a problem to meet the weekly wages bill of the twelve employees. Things gradually improved as work came in, and the partnership continued happily until the eve of Hitler's coming to power, on which day Rosenberger decided to emigrate, feeling rightly that Germany would no longer be a comfortable home for Jews, however keen they might be on motor racing. He left Stuttgart on 30 January 1933, and travelled to New York via Paris, to begin life again in the New World and under a new name.

His choice of Rosenberger as a partner shows that Porsche was by no means an anti-Semite, and he was soon to make clear, in his relations with the authorities of the Third Reich, that he was no respecter of persons, uniforms or political ideas. He had, in fact, little use for ideas as such, except in the workshop, laboratory and drawing-office. If one had to grade him by the height of his brow, he would come out as middlebrow, inclining sharply to low. He loved doing things with his hands, and he thoroughly enjoyed his motoring. When his son Ferdinand was a small boy the Doctor built him a miniature motorcar powered with a lawnmower engine, ostensibly for use on the private roads of the factory; and no one was less surprised than he when young Ferry escaped with it on to the road, where an indulgent police force looked the other way. At weekends, the Porsche family would go off to the lakes and mountains where Porsche had a shooting-box, and where he kept a fast motor-cruiser of his own design, complete with Austro-Daimler engine. There was also a privately-made, go-anywhere vehicle for shooting parties, with two seats in the front and one behind, with stowage for guns, rods and camping gear on each side. He also enjoyed sailing, and when ashore or off duty in town, he liked to drink beer with his friends in the cafés. Occasionally he would take in a movie or music-hall, but he never went to the theatre or to the opera. He would not read 'serious' books, and he detested controversial matters outside engineering. At the same time he was shrewd enough to realise that politicians had their uses. It was they who laid down research programmes and dished out the contracts, whatever a self-respecting engineer might think of them in private. Porsche was to see much of the Nazi politicos in the next twelve years, but he refused to be daunted by the *Partei* mystique. When almost every German male had a uniform of some sort, Ferdinand Porsche wore plain clothes, sometimes to the scandal of Hitler's entourage—as when the Doctor turned up at a top-level tank demonstration attended by the Fuehrer wearing a civilian blue overcoat and an Anthony Eden hat that hailed unmistakably from St James's Street, London. What is more, he always said 'Good Morning' instead of *Heil Hitler,* to everyone, including the leader himself, and addressed him not as *Mein Fuehrer* but simply as *Herr Hitler.* It must have been pretty to hear.

The first design project at the Porsche Büro was a new car for Steyr-Wanderer, although to remove any apprehension on the part of his clients he gave this design the number 7 instead of 1. The Doctor could be a politician himself when it suited him. He had actually begun on this design before he opened the Büro, having been in negotiation with Wanderer. This car, which afterwards became the 2 litre Wanderer, had an 1,860cc ohv engine using wet cast-iron liners in an alloy block. He used a beam axle at the front but swing-axles and radius rods for the rear end, and the suspension was by transverse leaf springs both fore and aft. Number 8 followed swiftly and was more exciting, being destined for Porsche's personal use. This was a 3·25 litre straight-eight with streamlined coupé body, and the engine was very soon developed to take a Roots blower, permanently engaged, as the heart of quite a potent sports car.

At this time the Auto Union combine was formed by merging the firms of Horch, Wanderer, Audi and DKW, a union symbolised by the trademark of the new concern, which was four interlaced circles, rather like the badge of the Olympic Games. The commercial side was looked after by the old-established firm of Phänomen, and Porsche set about designing them a lorry. He was also busy on a swing-axle for Horch and an overdrive for Wanderer. This was in 1932. He also became preoccupied with springing. Transverse leaf springs are not elegant engineering, and what is more they are heavy. The solution he arrived at was torsion bars and a trailing link, an arrangement later famous as 'Porsche ifs'. (See drawing, p 184.)

The torsion bar was not, of course, Porsche's invention. This arrangement, which can be described as simply a coil spring that has been unwound, had been used for many years on door hinges and is in fact the classic form of spring hinge for cigarette cases. It was not even new to the motorcar world, as there was a twenty-year-old French patent that had not been proceeded with, and Parry Thomas had employed a torsional auxiliary spring on his Thomas Special racing cars at Brooklands at least seven years before Porsche. But the Porsche trailing link was both ingenious and new, and so was his system of stowing the torsion bars inside a weatherproof tube running athwart the chassis. Moreover, by placing a second trailing link above and parallel to the first, pivoted to the frame and combined with a shock-absorber, he achieved a parallelogram linkage that would hold the wheel always upright, and prevent the variations in roll (and consequently in slip angle of the tyres) which are a disadvantage with a swing-axle design. There was another nicety of installation: each torsion-bar was splined at both ends, but with more splines at the outer than at the inner end, thus providing a sort of vernier adjustment. Supposing the outer end had thirty splines, the inner or fixed end would have twenty-nine. Thus if the whole system were moved 'one tooth', the outer end of the bar would be turned through one-thirtieth of its circumference (12 degrees) while the inner end anchored to the chassis would have moved through one-twenty-ninth, or 12·4 degrees. Thus it was easy to preload the torsion bar supporting any corner of the car to compensate for variations in height or load.

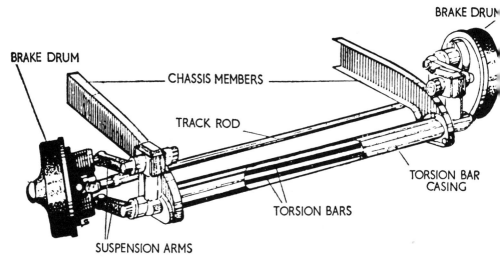

'System Porsche': independent front suspension by torsion bars housed inside a cross member

The first vehicle to profit by Porsche's discovery was known in the Büro as Type 17, and to the outside world as the 2 litre Wanderer, which appeared with Porsche front end in 1932. There was an immediate clamour from motor manufacturers all over Europe, who wished either to use torsion bars under licence, or wanted Porsche to design them a complete front end. Among the names we find Alfa Romeo, Austro-Fiat, Citroën, Delaunay-Belleville, Hanomag, Morris, NSU, Standard, Vauxhall and Volvo. Others decided to wait until 1950 when the patent would run out.

Designs for a racing car with ifs were probably already taking shape in Porsche's mind at this time, as were plans for a small car for the masses, but before either an Auto Union racer or a Volkswagen could proceed very far there was to be a considerable interruption. Dr Porsche was invited to Russia by Stalin, who wanted him to 'advise on industrial matters'. What he really wanted, it turned out, was for Porsche to inspect the whole country with an eye to its industrial future, and to become a sort of Minister for Technology. He was to design cars, tanks, aero-engines, tractors and the rest. Porsche said he would see, and see he did. They showed him, if not everything, at least a good deal more than any other foreigner had been shown. He went behind the Urals to inspect the great new factory towns and the hydro-electric plants, he visited car factories, foundries, and engineering works of every sort. The offer was very tempting. Porsche was to settle in Russia permanently with his family, and be given the fullest VIP treatment. What is more, he was to have a free hand technically to do what he liked. He travelled in a luxury train with private Pullman all complete, and it was the same when he arrived in Russia. Every luxury, and splendid technical liaison. But Porsche was not born to live in a rarefied atmosphere. He wanted to get out in the shops and meet the workers—an unheard-of thing apparently in a People's Democracy. He was living, he felt, in a gilded cage, and a most alien cage at that. He decided to remain in Stuttgart where his friends and his interests were. Of the marvels he had seen he never spoke, and although during the war he was working closely with Hitler and the various service ministries, who were no doubt burning to know what equipment Russia had been developing at the time of his visit, he always respected professional confidences. And to their credit, the authorities never pressed him, although when during the campaign of 1941 they discovered that Russia possessed presses that could form panels from four-inch armour plate, a thing Germany had no equipment for doing, they must have longed to pick Porsche's brains.

Probably another consideration that weighed with Dr Porsche was the lack of motor sport in Russia. All his life he had been concerned with motoring competition of one sort or another and it would have been hard for him to abandon it completely. Plans for a new GP formula to put an end to the period of *formule libre* anarchy were announced after the Paris Salon in the autumn of 1932. This formula was to run from 1934 until the end of 1936, and like all previous (and subsequent) sets of Grand Prix regulations it was designed to limit the speed of cars which everyone agreed had grown undriveably fast. There were to be no restrictions on superchargers, engine size, fuel and the like; simply a *maximum* weight for the complete car, without fuel, oil and water, and without its tyres (which vary in size from circuit to circuit) of 750 kilogrammes, or roughly 14½ cwt. This, thought the wiseacres in Paris, would limit constructors to between two and three litres of engine, and some 250bhp. Porsche saw things differently; it would obviously be a frightful risk for the little new Büro to set about designing a full-scale Grand Prix

L

car of futuristic layout without a paying customer in sight, but Rosenberger was as keen as Porsche, and made only the sensible stipulation that a separate company should be formed. Preliminary engine sketches were made on 1 November 1932, a new subsidiary was registered six days later, and on the fifteenth a meeting was held to decide upon the lines of the new car.

The way in which he approached the design was typical of Porsche. As this was uniquely a weight formula, he discarded all the things that tend to make conventional motorcars heavy, beginning with the propeller-shaft, which he abolished by placing the engine, gearbox and final-drive mechanism in one bloc at the rear of the car, with the engine nearly amidships, in front of the back axle, and the gearbox behind it. The driver sat right forward with the fuel tank behind him almost in the middle of the chassis where the difference between a full tank and an empty one would make little difference to the road-holding. The rear-engined layout was a bold one, as although it had been done before, by Dr Rumpler in his *Tropfwagen* Benz of 1923, that model

Dr Porsche's original specification for the 'P-Wagen' GP Auto Union, dated 15 November 1932

had not been a great success. The soundness of his thinking is shown by the fact that today, almost forty years after those first few lines appeared on the drawing-boards of Porsche and Rabe, there is not a GP car made that does not have the engine behind, in the fashion popularised by the 'P-Wagen' of 1932.

Looking back upon the 'P-Wagen', it is no longer the rear-engined chassis that strikes one as daring, nor the Porsche front-end with its trailing links and torsion-bar suspension, nor yet the swing-axle at the rear sprung first on transverse leaves and afterwards on torsion bars; all these things have become commonplace—thanks, be it said, to Dr Porsche. The really unorthodox feature resided in the engine. One might have expected Porsche to plump for overhead camshafts, as he had done in the Austro-Daimler and Mercedes days; instead he chose an ingenious pushrod arrangement, in which a single camshaft lying in the vee of the V-16 engine was used to work two valves per cylinder inclined at 90 degrees to one another in a hemispherical head. This was clearly lighter than the only viable alternative of twin overhead camshafts to each block, although it might be expected to limit the maximum revs. Probably indeed, the sixteen-cylinder layout was chosen not to encourage high engine speeds as one might imagine, but for two other very good reasons, both of which were in Porsche's mind from the outset: each cylinder being so small the reciprocating parts could be made very light; and by the same token it would be easy when the need arose for greater power to increase both bore and stroke without much altering the overall dimensions of the engine.

The original size of the R-type engine, as it was called in the drawing-office, was 68 × 75mm—figures later familiar as those of the 1100 Fiat and other successful designs. This gave just over 272cc per cylinder and a total cubic capacity of 4·4 litres, and the original version was expected to turn at 4,500rpm, rising to a maximum of 6,000rpm later in the engine's development life. The aim was always torque rather than revs, and anyone who watched Hans Stuck in action with the big short-chassis car when he broke the record for the Grossglockner hill-climb will retain an impression of Buick-like flexibility and immense power accompanied only by a low rumbling sound. This special car had one of the last of the V-16 engines, made when the bore and stroke had increased to 77 and 85mm respectively and the power was 545bhp from 6·33 litres at 5,000rpm— or possibly more in the sprint engines. Another advantage of the multi-cylinder layout was the low piston speed made possible by the short stroke, for even with 85mm stroke this was only 2,750rpm. The table on p 188 shows the development of the Auto Union racing engines.

	1934	1935	1936	1937
Capacity cc	4,360	4,950	6,010	6,330
Bore/stroke mm	68/75	72·5/75	75/85	77/85
Compression ratio	7·0:1	8·95:1	9·2:1	9·2:1
Maximum rpm	4,500	4,800	5,000	5,000
Bhp	295	375	520	545

By looking at this, however, we are peering into the future. When Porsche drew the first lines he had no customer in view, although he may have heard rumours that Horch, Audi, Wanderer and DKW were going to merge. He may also have foreseen the rise of the National Socialist party to power, and known of Hitler's extreme fondness for motor racing; be that as it may, the day came when someone at Auto Union's thought that racing would be a good way of popularising the new name. Did Dr Porsche think he could design them a suitable engine? 'I have already done so', said the Doctor, speaking for once in a Jeeves-like voice. Thus, when the new government gave support to German racing at the start of the 750 kilo formula, there were two contenders in the field: the Auto Union 'P-Wagen' plus a straight-eight twin-camshaft Mercedes-Benz (the Type 125) with independent suspension all round. This design, brought on by Dr Hans Nibel and Herr Wagner, was conspicuously in the Mercedes tradition—and by no means unlike the design for a twin-cam racing 3 litre that Porsche had left behind him when he moved away from the firm.

Besides the racing cars, the change of government was to make possible the fulfilment of another project on which Porsche had been working for a number of years, namely a People's Car. In this project Porsche was about two years ahead of Hitler, for as early as 1931 the Büro had drawn up plans for Type 12—their fifth undertaking since numbering started at number 7. This was a truly prophetic vehicle containing almost all the elements of the ultimate design: beetle-shaped 'streamline' body, engine at the back, independent suspension all round, a central backbone frame, and the spare wheel at the front beneath an 'alligator' bonnet. This car was to be built and developed by Zündapp, the well-known motorcycle people, and it was they who asked for a five-cylinder water-cooled radial engine. They hoped that entering the motorcar business would assist their recovery from the recent slump,

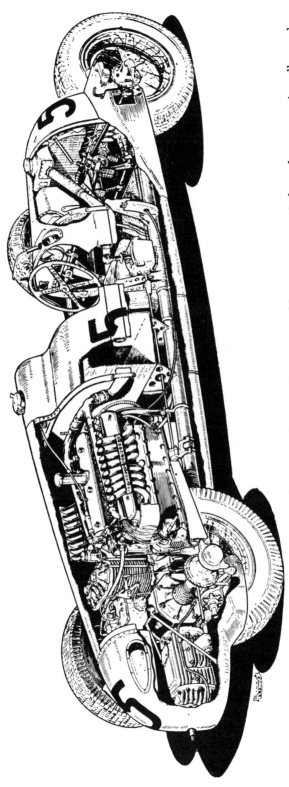

The 1936 6 litre Auto Union Grand Prix car: V–16 cylinder central engine, 520bhp at 5,000rpm, independent suspension all round

which had hit the motorcycle trade very hard. The car was road tested
in 1932, but by this time the slump was over and Zündapp required all
their factory space to build motorcycles. Porsche received a fee of
85,000 marks for his work, and that was that. Meanwhile Porsche was
approached by another motorcycle firm, this time NSU, who wanted a
1½ litre car.

The Doctor's answer was named Type 32, and differed from the
Zündapp in having torsion bars instead of transverse leaf springs for
the suspension, and what the Germans call a 'Boxermotor'—a flat-four
with air cooling assisted by a fan. In choosing this engine form, Porsche
was not merely anticipating the eventual VW of today, he was also
harking back to the design of a 90hp aero-engine he had constructed in
1912. This, too, had overhead valves by pushrod, and finned air-cooled
cylinder barrels, but instead of being arranged as a pair of flat twins side
by side, the opposed cylinders were arranged as a pair of narrow-V
twins, no doubt to arrive at the shortest possible four-cylinder engine.
Three prototypes of the Type 32 car were built, and proved very
promising. There was slightly more room than in the present-day VW,
and with 28bhp for cruising or 30 at peak revs, the machine had a maximum speed of over 70mph. But NSU backed out as Zündapp had done,
and for similar reasons—the investment was too great.

None the less it seemed to Porsche that the new government, with
its great road plans and obviously sympathetic approach to motoring
matters, would certainly look favourably on the idea of a People's Car.
So he prepared a memorandum on the subject for the Minister of Transport at the end of 1933, and submitted it on 17 January 1934, just as
similar plans were taking shape in Hitler's mind. Here an old friend,
Werlin, late director at Mercedes-Benz, pulled some strings; Porsche
obtained an interview with the Fuehrer, and submitted his famous
memorandum.

This well fitted the mood of the times. He referred to the recent
'people's radio set', which had vastly increased the listening public, and
by so doing had created a new industry and a new kind of retail trade.
Radio shops sprang up everywhere, and each of these required a more
or less trained radio mechanic. A 'people's car' would have the same effect
on a larger scale, and its production would suit Schachtian economic
theories very well, for a man who was spending money on a domestically
produced motorcar would have that much less to spend on goods bought
with Germany's scarce foreign exchange. The government would find
it a convenient way of 'channelling demand'. Porsche then put up his
proposition.

It was no good simply scaling down some existing design. A people's car must be started from scratch—and he sketched out something very similar in layout and performance to the Volkswagen as later produced. If the government would entrust his Büro with the job, he would guarantee to produce and develop prototypes within twelve months. If, after due trial, his car was considered satisfactory, the government could place orders with the industry for its production. Porsche offered to start right away and work without fee; on condition that the Büro's out-of-pocket expenses on the project were reimbursed by the government, and that if the model were eventually put into production the Doctor should receive an agreed sum per unit for the use of any of his own patents embodied in it. Porsche then permitted himself a personal testimonial, in which he declared that, during his long career in the motor industry, he had been responsible for the development of sixty models, and so knew what he was talking about; his colleagues were all fully-trained men and imbued with his own ideas, and furthermore, the Büro was perfectly independent, having no ties with the industry or trade. Finally, he drew attention to the new 'P-Wagen' racing car which had just been built with the help of a government grant, and had made a splendid début with Hans Stuck when it went record-breaking on the Avus.

Initially, there were two rival designs, based respectively on the still-born Zündapp and NSU, except that the latter was to have four speeds and a shorter stroke, and the former's radial engine was exchanged for a three-cylinder two-stroke. The idea of a two-stroke appealed to him because of the small number of moving parts—a great consideration in a car for the marginal motorist; in fact, when work started, he simplified even further and replaced the 1,000cc radial by a parallel twin of only 850cc. This was announced in a progress report dated 31 January 1935, where he stated that three types of engine were being studied, of which the most promising from the point of view of weight and price seemed to be a 2/4 cylinder opposed-piston two-stroke with crankcase compression, for which foundry patterns already existed. This, then, was the first engine fitted to the Volkswagen, but when development snags held up progress, Porsche decided to drop the two-stroke in favour of the air-cooled flat-four. He just had not the time, nor the resources, to 'get the bugs out' of a new two-stroke.

For naturally, notwithstanding the official promise of 'all the assistance necessary from government and industrial sources', Porsche was in for a hard time, with much jealousy and obstructionism to contend with. The 'total resources' promised amounted in actual fact to a grant of 20,000 marks a month for ten months, or about £16,500 in all, Porsche to supply premises, equipment and personnel. These, for the record,

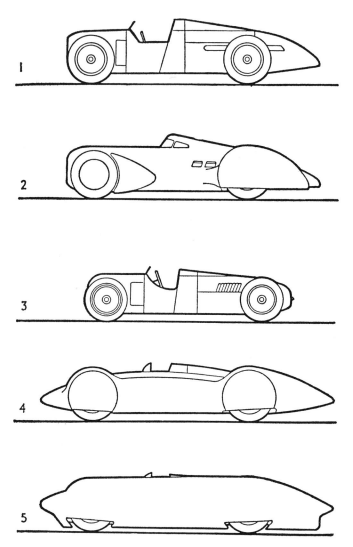

Auto Union competition cars. (1) The original Auto Union Grand Prix car, Avusrennen, 1934. (2) The spatted 'saloon' car with which Hans Stück broke several world records in February 1935. (3) The 6·1 litre car in which Bernd Rosemeyer won the German Grand Prix of 1936. (4) The fully-streamlined Auto Union record-breaker of 1937. (5) Elevation of the car in which Rosemeyer was killed during a record-breaking attempt in January 1938

consisted of the garage at Porsche's home containing two lathes (one small, one medium-sized), one milling machine, one boring-machine and two or three lesser machine tools. The labour force comprised twelve men, including Porsche's assistants in the drawing-office. Everything, in short, had to be bought out, any alterations and hand-fitting being carried out in the workshop. These were the conditions under which the first three test cars were made which sallied forth from the Villa Porsche during the autumn and winter of 1936, covering 30,000 miles each, driving day and night on a tough duplex circuit comprising one-half *autobahn* and one-half hilly going on the twisty roads of the Black Forest. Sometimes the driving was shared by official observers and experts from the motor industry, at the end of which test the observers prepared a report—and although this was couched in the most patronising terms, they could not withhold a grudging approval. The word was given to go ahead.

The original series of 'VW3' having proved a success, a further series of thirty cars ('VW 30') was ordered, and these were built by Mercedes-Benz, to be followed shortly by a second batch of thirty. It is interesting to recall that the short-lived rear-engined 140 H and 170 H (*Heckmotor*) Mercedes came out in the same year and looked, from the front, very much like the Volkswagen prototypes. There followed a prolonged series of tests over the same ground during which one car logged 62,000 miles of uninterrupted day-and-night running, and all cars carried devices to record the number of times clutch, brakes and gear-shift were used. What is more, the drivers during this extended trial were not engineers but members of the SS who happened to possess driving licences. In this way the cars got the sort of beating they might have received from the public (although it is not recorded that women drivers were given a chance) and the security so beloved of the Nazi regime was preserved. By this time the State had taken charge of the project, which henceforth would be controlled by a government-owned Volkswagen Development Corporation. Shortly afterwards (1938), the name of the car was officially changed to 'KdF-Wagen', from the Nazi slogan 'Power through Joy'. At the Berlin Motor Show of 1938 Hitler was able to announce that now, after four years' intensive development to improve the product and reduce the price, the People's Car was about to go into production. The foundation stone of the factory was laid three months later, on 26 May 1938. The selling price was to be 990 marks, which, although 10 per cent more than that decreed by Hitler four years before, was little more than £80.

The KdF of those days still had the 995cc engine used in the prototypes, the larger engine fitted in post-war VWs being a development of the military version which appeared in many forms (but relatively small quantity) during the war.

It was once fashionable to decry the whole Volkswagen project as a gigantic confidence trick, played on the German public by the Nazis. The timetable does not support this view. Four years is by no means an unreasonable time for the design, testing and production planning of a brand-new model, especially one of such novel conception. Advance publicity made the time seem longer than it was; but no doubt this barrage of blurbs was needed for two reasons. In the first place a huge guaranteed demand was needed to finance the building of the factory and get the price down, and in the second place one of the main aims of the project was to siphon public savings off the retail market, and thus combat the inflation that always accompanies a programme of public works and military preparation. The factory did in fact start producing vehicles—civilian ones, for the military version did not appear until later—before the end of 1939, and there is little doubt that had not Hitler invaded Poland, early VW subscribers would have been getting their cars, for the target production figures were 800,000 to 1,000,000 cars a year. As it was, the Volkswagen factory was given other munitions work to do, and although several military versions of the VW were made, including a cross-country vehicle, an amphibian and an ambulance, only 100,000 in all were produced between 1939 and 1945.

In the five years before the war Porsche continued to work with bewildering versatility. At one stage, preoccupied with the problem of vanishing coal and oil resources, he did a lot of work on latter-day windmills, applying modern aerodynamic theory to the design of the propeller blades. Results were quite encouraging, and he produced units delivering from 1 to 10 kilowatts, the energy being stored in big accumulators for domestic or other use. He also worked hard on tractor design to improve the state of Germany's rather backward agricultural economy. Towards the end of the period, he went back to his beloved Steyr 'Austria' design, and produced an americanate saloon that was promising well when war stopped play.

Various kinds of munitions work also kept him busy, including a stationary engine for use on searchlight and AA gun sites, developed from the VW. Then there were a bus chassis with five-cylinder, under-floor radial engine; a new system of rubber suspension; an opposed-piston two-stroke; a petrol-electric car heater; a sprung rear fork for motorcycles, and an astonishing $1\frac{1}{2}$ litre racing and sports engine which was arranged as a V-10. There was also talk, when the original 'P-Wagen' was on the stocks, of a sports car using the same $4\frac{1}{2}$ litre V-16 engine either blown or unblown, with engine amidships as in the racing car. This was to have been a three-seater, with the driver sitting between his two passengers, as in certain Panhard models of the period, and in a post-war design proposed by Jean-Pierre Wimille. There was also the great car designed for attacking the Land Speed Record.

This record, the most glamorous of them all, had become a British monopoly through the efforts of Sir Henry Segrave, Sir Malcolm Campbell, George Eyston and John Cobb. It was natural that Germany should want to take a hand, and equally natural that she should turn to Porsche for the design and to Mercedes-Benz for the construction of a suitable car. The Doctor signed on as special consultant, and discussions began in July 1937, when Eyston was building 'Thunderbolt', the car that was to raise the record to 312mph at Bonneville, Utah, shortly before Christmas of that year.

Porsche's task was immediately bedevilled by one of those politico-sentimental decisions so frequent in dictator countries: Korpsfuehrer Huehnlein, head of the Nazi directorate for motoring sport, decreed that a German car might break records only on German soil. When it was pointed out that no suitable stretch existed, he said 'Then we'll build one'. The result was the Dessau Rekordstrip, which was in fact merely a six-mile section of *autobahn* of rather more than twice the normal width, being built without a dividing centre strip to give just under 60ft of carriageway. This was trifling both in length and breadth compared with the 14 mile straightaway in Utah and the almost un-limited spread of the salt lake, and it set some very special problems in acceleration, braking and stability. Porsche calculated that to reach a speed of 600kph (372mph) the car would need a run-in of 6,300yd before the measured kilometre, and it would then take at least 2,400yd to pull up, unless special air-brakes could be developed in time. Further-more, if a measured mile were to be included, as it would have to be to derive full benefit from publicity in Britain and the United States, the braking distance would be shorter still, and the driver would find him-self with less than a quarter of a mile in hand before a corner put an end to the 'recordstrip'. Porsche thought this was crazy; and he also had his sights upon a speed of 700kph (435mph), which he thought his car might attain in favourable conditions at full revs. The attempt was scheduled to take place during a Dessau record week in 1940. It is probably fortunate for all concerned that it did not come off.

Probably the car could have done it. In appearance, Porsche's monster was not unlike the fully-streamlined record-breaking Auto Union, having an enclosed streamline fairing round the driver's head and humps over the wheels; but no doubt with Rosemeyer's death in mind, there were horizontal fins like a miniature aeroplane wing on each side, set at a negative angle of incidence to keep the nose from rising—an interesting anticipation of modern GP practice. An inverted V-12 engine was to be used, developed from the DB 601 unit which was to make such a name in fighters during the war, and rated at 2,500bhp, with a maximum of 3,030bhp for sprint use. This 44 litre engine was placed amidships behind the driver (Auto Union again), and there were six wheels, on three axles: normal steering in front, and four-wheel drive to the pair of rear axles. Weight was kept down to 2·8 tons, and Continental produced special 7·00/32 tyres said to be good for 700kph. There was no gearbox, as the car would be pushed up to a speed at which the clutch could be let in. The car measured some 27ft overall, and the light-alloy skin of the detachable body was only from ·012 to ·040in thick. Patents were granted in the joint names of Dr Porsche and his slide-rule man, Mickl, and although the project remained a State secret in Germany, a leak occurred somehow so that an accurate description of the car appeared in the English motoring monthly, *Speed*.

During the war, all sorts of designs poured from the Porsche drawing-boards: modernised windmills, tractors with hydraulic attachments, military vehicles in profusion, including the 'Mouse'—a sort of mechanised 'Castle Horrible' weighing 180 tons, built in accordance with a Hitlerian whim.

*   *   *

It is not surprising, perhaps, that when the war stopped, the Allies should have arrested Dr Porsche. He had, after all, been extremely close to the Fuehrer, and his ideas had greatly furthered the war effort. The Americans, who first took possession of the old Doctor, now rising seventy years of age, treated him relatively well, recognising him as a technician, not a Nazi politico. But political boundaries and political complexions change. Dr Porsche was transferred to French keeping, and life was less pleasant. Taken to Paris and housed uncomfortably in the porter's lodge of Louis Renault's town house, he was consulted on the development of the 4CV, the rear-engined layout of which may have owed something to VW thinking. Later he was transferred to Dijon, but it was not until his captivity had lasted almost two years, from 27 October 1945 until 5 August 1947, that he was finally released, bought back from the French, it is said, with money paid to the Porsche Büro for designing the flat-12, four-wheel drive Grand Prix car for Cisitalia. French friends in the racing world, including Charles Faroux and Raymond Sommer, also interceded on his behalf.

Returned from captivity, the old Doctor gave his approval to the GP design, murmuring (we are told) 'I could not have done better myself, in fact I would not change a line.' Official Porsche history is a little vague, one might almost say obscurantist, concerning the Doctor's own part in designing the Porsche 356 sports car; it is admitted that there were plans for a sporting machine based on the Volkswagen before the war, but apparently the National Socialist Workers' Front would not allow it to be built. Perhaps they were right. The Doctor was always one to scatter his fire, and had he been allowed to start on a sports car then, other valuable work might have suffered. Be that as it may, Ferdinand Porsche came home in August 1947, and lived to see motorcars bearing his name win high honour in the competition world. Around him were most of his old and trusted associates, including Rabe, the chief engineer from Austro-Daimler days, Mickl, the aerodynamicist, Hrushka, Kommenda, the body designer, and Kales. The Doctor's seventy-fifth birthday was celebrated by a rally to the works of Porsche owners from all over the world. Shortly afterwards he suffered a stroke, and died, on 30 January 1952, in his seventy-fifth year.

Whatever part he played in the last few years of his life, after the hardships of exile and imprisonment had reduced his vigour, there is no doubt that Ferdinand Porsche remained an engineer to the last, the greatest and most versatile genius the motorcar has produced.

7 HARRY MILLER

by GRIFFITH BORGESON

HARRY MILLER (1875–1943). American: The USA's greatest and most successful racing-car designer between the wars. Built chassis with conventional layouts, front-wheel drive, four-wheel drive and all-independent suspension. Supercharged racing engines based on Miller's designs are still in use at Indianapolis in 1970.

HARRY MILLER

IT IS NO exaggeration to say that Harry Miller is the greatest single figure in American motor-racing history, as well as one of the greatest automotive designers that the United States will ever produce. The evidence is in his abundant achievements, up to and including those engines called Offenhausers and their decades of inexorable leadership, and in his immortal influence upon American racing machinery, He made several fortunes by building the finest racing equipment in the New World and he lost them the same way. As one of his life-long associates has said, 'Harry never held back a penny on research or experiment. He was always ready to gamble his last dollar on the drawing-board. His whole career was a single-handed crusade devoted to mechanical perfection.'

In his heyday, Miller stood five feet six inches tall, had black hair, blue eyes, a fair complexion, ruddy cheeks, a medium build and a slight paunch. He favoured grey suits, black shoes, *boutonnières*, a Paul Whiteman moustache and expensive hats worn at a jaunty angle. He put a lot of the roar into the Roaring Twenties and he looked the part. He presented an impression of easy affluence, lively good taste and debonair conviviality mingled with dignified reticence.

Of course, Miller's affluence was actually extremely variable. He came from a very humble background and repeatedly worked his way through cycles of poverty, wealth and back to poverty again. He never showed any reaction to the outgoing of money in any form, including total, abject bankruptcy. 'Oh, I'll make some more of the stuff', was his stock, indifferent comment.

His last fortune seems to have exceeded a quarter of a million dollars, but when his wife would ask him for just a little bit to put away for a rainy day he would say, 'Aw, Honey, I make it and I'm going to spend it. As long as you have everything you want and enough to run the house on, that's all right.' When a close friend suggested that Harry should put his Malibu ranch in Edna's name to protect it from foreclosure he snapped, 'I wouldn't think of it. She might throw me out.' This was the sheerest fantasy, but real enough for him since he trusted no one. He was the most inept of businessmen but would trust no one else with the most crucial, profit-and-loss aspects of his business.

Miller's good taste was congenital and he was definitely more artist than engineer. He was the originator of an artistic approach to racing machinery which became an American tradition, but he carried it to a fanatical extreme. When parts of his machines broke because they were too light or frail, he never complained and methodically took corrective measures. But when a part proved to be heavier than need be, the worker responsible for it could count on the full blast of The Old Man's outraged anger. Each of his cars took between 6,000 and 6,500 man-hours to build. Of those, about 1,000 hours were invested solely in beautifying the machines, in hand-finishing and polishing each part.

Although it was quick to pass, Miller had a very sharp temper. He was abruptly frank in his personal relationships and if he did not like a person he would promptly tell him so to his face. He was a hard man to work for and expected his men to match his own tireless energy that would let him work for seventy-two hours at a stretch. He thought nothing whatsoever of risking payroll money on his own personal projects and letting his men wait for what they had earned. Many, many of them quit in anger. But most of them soon remembered his benevolence, his warm affection towards the whole crew, and the thrill and satisfaction of being part of it. Most of them hurried right back to The Old Man knowing that he would welcome them and never mention their absence.

Miller's conviviality was hearty and genuine providing you could speak his language, which dealt almost exclusively with racing cars and engines. On nearly all other subjects he was practically speechless. He once walked out on a millionaire would-be client muttering, 'That man talks engines like a fool!' But among 'the boys', among those who understood and shared his own mystique, he was garrulous, affectionate and fun-loving to the point of pure boyish mischief. He always loved animals, above all those that he found amusing. He kept dogs and named them after his favourite men in the game, like the mutt 'Rick' whom he named after Eddie Rickenbaker, the then-boss of the Indianapolis Speedway. He always had monkeys around him. He used to drive to the plant with one in his car and loved to see it make a mess of someone's work in the shop. This vicarious irresponsibility, even sabotage, co-existed with his compulsion towards perfection. He kept a parrot which he used to let fly freely in his 70ft long drafting-room. It would cackle, 'Damned Watts Local!' every time the Pacific Electric train would rattle the building's windows. Its fondest sport—which Miller loved—was zooming down to peck Fred Offenhauser's bald head whenever that hard-boiled little foreman came in from the shop with a serious problem. Leo Goossen still owns the crusty old parrot.

Miller did not run his firm in a businesslike manner, even though it reached big proportions in the late 1920s. You could send him an order for parts and perhaps not get delivery for three months in spite of all your heavy and costly racing commitments. At the same time, Miller would extend credit to anyone, even to people whom he knew never would pay. 'We've got to keep the show going', he would say to an irate Offenhauser who was somewhat more concerned with keeping the business going. When things became really tough for him Miller borrowed, even from his few employees who could afford this luxury, and as his ability to meet payrolls dwindled away, so did his staff. At Indianapolis in 1934, he ran into one of his old favourites from the plant. Embracing him like a long-lost son Miller said, 'Walt, one of these days I'm coming back to LA and pay all you boys every cent I owe you.' Then tears began pouring down his cheeks and he brusquely walked away. Walt Zabraske, still a top man in the Offy factory, never saw The Old Man again.

Miller's dignified reticence was really a cover for intense shyness. It was crippling. As America's greatest designer of racing machinery, he was constantly asked to address civic and professional groups. His eternal answer was a flat and accurate, '*Impossible!*' The shyness stemmed from his lack of academically acquired knowledge; he had run away from home as a child and his schooling had stopped at the primary level. His 'engineering drawings' were mere scrawls and he was ill-informed on many subjects that were close to his art. He depended heavily upon his men to guide him in these spheres, and as long as he had this guidance his brilliant ideas found easy expression. But no one ever boasted of helping The Old Man. The privilege of working for and with this great talent was a profound satisfaction in itself.

Miller never showed the slightest reaction to failure. A man could come to him and say, 'Harry, I've blown up that engine on the test stand.'

'How bad?'

'It's totalled.'

'Well,' Miller would say, purposefully and philosophically, 'let's see how fast we can build another one. Get going.'

Miller was passionately intolerant of any meddling with his completed designs. Valve stems and connecting rods in the 91cu in engine were points of failure to which The Old Man was perfectly resigned. When he learned that Frank Lockhart, who was campaigning one of the works cars in the East, was designing his own valves and rods he snorted, 'Why does every kid who wins a race suddenly think he's an engineer?'

He wired Lockhart saying 'FORBID MODIFICATIONS.' Lockhart wired back, 'HOW MUCH DO YOU WANT FOR THE CAR?' He bought the car, made his changes and won the National Championship with the fastest and most reliable '91' by far.

Miller was extremely successful in his heyday but could have been infinitely more so had he had just a little patience. It was typical of the man that he lost interest in most of his projects long before they ever reached completion. 'Look,' he would say to an assistant, speaking of an intricate, half-completed design, 'this is what I have in mind. You take it over and finish it.' And then Miller would go charging off into the new unknown.

When Miller is remembered at all today it is for a very few of his creations, but there were scores upon scores of them, all brilliant and often never finished or never properly developed. His greatest success was with the '91' which he built for the 1926-9 formula. He built about fifty of these engines and thus was able to develop them to a high degree of refinement, which is not possible with engines built in lots of from one-half to a handful.

Harry was a 'funny' man, his wife used to say, meaning strange rather than amusing. He required little sleep and would often work for three days and nights straight through. At home, he would sit and think, or lie awake and think, for hours at a time as though in another world. This contrasted sharply with his behaviour when the meditative mood was not upon him. Then he was all bright spirit, affection and teasing prankishness.

Perhaps there was a fundamental rift during Harry's and Edna's early days together, when she used to be frightened by his habit of taking whole strings of sentences out of her mouth before she uttered them. She was frightened, too, by other aspects of this seeming clairvoyant flair of his. He would say, 'You know, so-and-so is going to die next Tuesday at about nine o'clock.' Not 'I *think* that . . .', but just the flat statement. After the third or fourth of these predictions turned out to be accurate, Edna asked him to keep them to himself and he sealed off this side of his personality from this happy and beautiful little woman.

Perhaps if she had been able to live with his hunches, she could have led him to the grand destiny that she always knew was his and wanted for him, but which he had no compulsion to achieve. Of course, he did not want to be another Henry Ford or Walter Chrysler, but there was something else he could have become if he had a life-partner who tried to understand the weird side of his genius as well as appreciating its results. But Harry and Edna communicated well on the level of beauty. The Old Man consciously, deliberately and without shyness regarded himself as an aesthete. He never completed a car in Los Angeles without proudly taking Edna to see it first, and it was partly to please her that he put so much beauty into his work.

In her own way, Edna Miller was basic to the Team, I am sure. So was Fred Offenhauser, with his cunning hands and literal mind and with the rapport that could exist between himself and Miller only on a very nuts-and-bolts level. But Miller was a strange and complex man who revealed different aspects of himself to different people. There was, for example, Leo Goossen, the lanky, sensitive kid of a draughtsman who came to work for him in 1920, supposedly with only a few years to live.

'Leo,' Miller would say, a hand clutching scraps of paper covered with da Vincian scrawls, 'Leo, listen to me and try to understand this. *I* don't do these things. I get help. Somebody is telling me what to do. I mean it, Leo. I *rely* on it.'

Goossen was not disturbed by this, whatever his reaction may have been. Being a trained technician and in Miller's constant presence, and constantly seeing the improbable, often magnificent ideas that erupted from Miller's brain and fingertips, he may have thought, 'Why not? It's wild however you look at it.'

There was another person to whom Miller confided that he had a 'control'. This began in the early 1920s and Eddie Offutt, like Goossen, seemed to find nothing strange in the claim. Offutt moved in and out of Miller's life until the end but never drifted far from it. He worshipped the man and his ideas, whatever their source. When Miller hit hard times in Detroit in the years just before his death Offutt, working long hours on government war work, took it as a privilege that he was able to give Miller another eight hours of his every working day.

Miller's end was poignantly tragic. He lost his Team when he went bankrupt in 1932 and went wandering East to build new fortunes on his lustrous reputation. He designed and built many splendid cars, cars always more exciting, more novel than the last. But they didn't work well and were plagued by ghastly luck. Somewhere between the Ouija board and the drawing-board, between the Team and the solitary Old Man, there were certain elements missing. Missing were skilled men who would not 'yes' him but who, like Leo and Fred and Eddie, could and would say with friendly frankness, 'Harry, that's *out*. You just can't machine it! Now, if we do it this way . . .'. It did not help when he cut himself off from Edna, too.

Harry came down with a severe case of diabetes in 1929. Then there was his bankruptcy and the sheriff's auction block and the fruit of his life's work scattered for a penny on the dollar. Then the years of struggle all over the East and their endless promises of success which always turned to dust. And finally Detroit.

Detroit never had had any fondness for Harry Miller because he was a source of embarrassment to the industry. When the legitimate industry's finest engines were offering the public all of 0·25hp per cu in, Miller's thoroughbreds were pulling around 2·75hp per cube. He found no haven, no employment even in a Detroit glutted with cost-plus war contracts. He had to take in job work to maintain a tiny shop. And there were other reasons why Detroit did not want him. He was old. He was a failure. And there was his face.

The last years of Miller's life were tortured by cancer of the face. It was a grim thing and he demanded that Edna remain in Los Angeles rather than be exposed to his ugly misery and squalid life. The growth was removed by surgery but he remained in constant pain and never left his room without a heavy dressing over the disfigurement. He weakened rapidly and his heart began to fail. When, in 1943, it was obvious that Harry was dying, Offutt telegraphed Edna. From a Monday to a Friday she moved heaven and earth to get on an airline flight but when she reached Harry he was dead.

There was Offutt, faithful to the last, to give his sympathetic help. There was a crazy, clowning gibbon in the poor little hole-in-the-wall shop. And there were the sixty-eight-year-old Old Man's newest inspired dreams, years ahead as always, on rolls of paper and in wooden patterns. And that was the end of the story.

*   *   *

Harry Armenius Miller was born in 1875 in Menomonie, Wisconsin. His mother was Canadian and his father was German (Meuller), a man who had studied for the priesthood but instead became a musician, painter and linguist. None of these proved to be a profitable vocation but he foresaw an ambitious business career for his son Harry. But of the three sons and two daughters in the family, Harry alone was gifted with mechanical aptitude and he followed this bent from an early age.

At fourteen, he was working in a local machine shop when the operator of the donkey-engine in a nearby lumber camp broke a leg. Miller filled in on the job, decided that the engine was a wreck and tore it down the first day. The foreman was aghast and threatened to kill him if he did not have it together and running in time for work the following morning. Miller lugged the worst parts back to the machine shop, refinished their surfaces and met the deadline. The engine was transformed for the better and the Harry Miller legend was on its way.

Seeking experience and opportunity he drifted to Minnesota, Oregon, Utah and then to San Francisco, where he worked in a bicycle shop. There he got his first experience with a petrol engine: one which was installed to run the lathe. He tinkered with it, learned some of its secrets, bought one of his own, installed it on a bike and had the town's first motorcycle. This was in 1896 and the legend has it that this was the first motorcycle in the United States. 'I guess I should have gotten a patent on it', Miller used to say.

He got married and the following year took his wife back to Menomonie to meet his family. He was always building something and here
he built a strange four-cylinder contraption which clamped on the back
of a rowing boat and propelled it. This seems to have been the first
outboard motor, at least in the United States. The idea did not impress
him particularly and again he neglected to patent it. Another young
machinist in Menomonie thought differently and soon built and
patented a two-cylinder outboard. He was Olie Evinrude, the father of
an industry.

Harry and Edna had already returned to the West Coast and, at
twenty-three, he invented *and* patented a new type of sparking-plug. He
sold the manufacturing rights and went on experimenting. At thirty, he
invented the Master Carburettor. It was an immediate success, its
popularity grew over the years, and in 1911 a new plant was built for
its manufacture in Indianapolis. That was the year of the first 500 mile
race and the beginning of Miller's active contact with automobile racing.

Miller conceived a better idea for a high-performance carburettor,
sold his interest in The Master and moved back to Los Angeles. There
he set up a good-sized plant to manufacture the Miller carburettor.

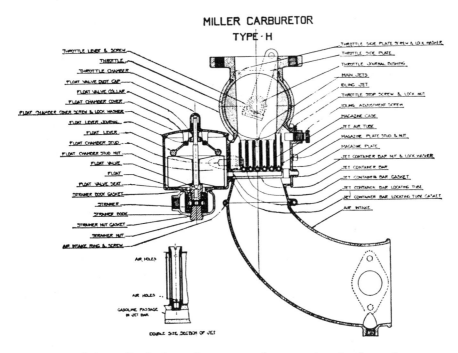

Drawing of the Miller high-performance carburettor showing how the rotary
barrel valve progressively exposed jets in characteristic Miller 'jet bar'

Nothing could touch it at the time and for years the plant turned out many thousands of units per month. (See drawing, p 207.) Passenger-car owners used them, and so did fire companies and police departments. They soon became used universally on racing cars and the racing fraternity came to make the Miller plant its headquarters and hang-out on the West Coast.

So in 1915 Bob Burman, America's King of Speed at the time, came to Miller with a new problem and challenge. His four-cylinder Peugeot had scattered its engine. There was no hope of getting a new one from France in time for the big Corona, California, Road Race which was just a few months away. Would Miller study the fragments and build him a new Peugeot engine, from the patterns on up? Miller conferred with Offenhauser, his shop foreman, got a laconic, 'Oh, yeah, we can do it', and found himself in the engine business. The new engine ran better than the old Peugeot but it, too, was reduced to rubble when Burman crashed fatally in that very Corona race.

In 1916, Miller designed and built his own four-cylinder engine. It used a barrel-type crankcase which was part of the aluminium casting which included the wet-linered block. The three main-bearing crank-shaft was made in two parts and had a single ball-race in the middle and double ones fore and aft. The head was detachable and a single camshaft, driven by a gear train at the front, actuated four valves per cylinder by means of rocker arms. The Peugeot influence was clear, although the engine bristled with originality. With it, Miller intro-duced aluminium pistons to America and established himself as a piston manufacturer. Six of these cleanly designed engines were built and one served well in a local doctor's aeroplane. Another powered Barney Oldfield's pioneer aerodynamic coupé, the 'Golden Submarine', a Miller creation in its entirety. Another went into a race car for the prominent driver Frank Elliott, who said in later years, 'It wouldn't pull the hat off your head'.

Now let's pause for a glance at the status of racing-engine design during this busy second decade of the century. In 1912, the team of Boillot, Gaux and Zuccarelli, with draughtsman Ernest Henri, had revolutionised racing car design with his Grand Prix Peugeot. For the first time in automotive history, inclined overhead valves and dual over-head camshafts had been combined in a way to extract the most from an engine's breathing potential. The engine was a shattering success and for 1913 the team added several refinements. One was replacement of the shaft-and-bevel cam drive by a train of spur gears. Another was adoption of a barrel-type crankcase, the sort which has no detachable sump. Its advantages are its own rigidity, plus 360 degree support of the main

Page 209 Harry Miller in his element

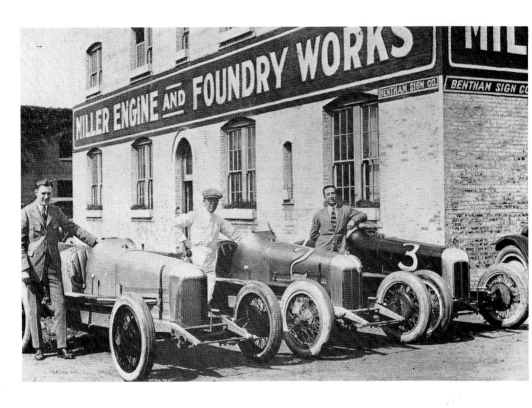

Page 210 (*above*) Miller 91s of the 1926–9 formula, outside the Miller works in Los Angeles; (*below*) Jimmy Murphy and Harry Miller with 122cu in 2 litre engine with normal carburation by four two-throat Miller carburettors

Page 211 (above) A V-16 Miller of about 1931: Bill White (left), Shorty Cantlon (in car), Henry Miller and "Doc" Cadwell; (below) Leon Duray in his '91' 1,500cc front-drive car which set an Indianapolis record of 124mph which was unbeaten for nine years.

Leon Duray-Miller Special

Page 212 (above) The front-drive Miller Ford of 1935, with Miller standing. The first Novi, essentially a V-8 Miller, was installed in one of these chassis; (below) Harry Miller's swansong: the highly sophisticated six-cylinder engine which he designed for the Gulf-Miller cars of the late 1930s

bearings. These are carried in large metal discs which are bolted to crank-case bulkheads after the shaft is fed into the case from the rear. This is the sort of engine that Burman had brought in a basket to Harry Miller.

In 1916, Bugatti had erected his famous sixteen-cylinder aero-engine. Instead of being a V-type it was a 'parallel eight', consisting of two eight-cylinder banks (two four-cylinder blocks each) mounted side by side on a common crankcase. Two crankshafts were used, tied into a central gear to which the propeller hub was assembled.

The straight-eight principle was not new but Bugatti gave it the most refined expression which it had received up to that time. To aid the needs of the Allied war effort the engine was built in series on both sides of the Atlantic. Ernest Henry worked for its Parisian constructors and in the USA the contract was awarded to the Duesenberg Motor Corporation, then located in Elizabeth, New Jersey. The Miller Carburettor Company was chosen by the US Government to provide carburettors and fuel pumps for the Duesenberg-built Bugatti and Miller, with Offenhauser, set up a small plant nearby for the manufacture of these units. Thus, Henry, Duesenberg and Miller all had the fullest exposure to the Bugatti straight-eight. They all worked with these engines and took part in their testing and development.

Instantly following the end of the war in November 1918, Henry joined Ballot for the purpose of designing a new racing car, specifically intended to repeat the pre-war French victories at Indianapolis. The Duesenberg brothers were equally swift in their action and plunged into the design and construction of their own in-line-eight racing cars. Through superhuman efforts Ballot and Duesenberg were able to confront each other at Indianapolis the following May, each with a team of revolutionary new machines. The victory went to a pre-war Grand Prix Peugeot but the new cars showed a sufficiently brilliant potential to set the engine-configuration pattern for the decade ahead.

Henri's Ballot engine borrowed little from Bugatti other than the straight-eight principle. Carried over from the old Peugeots were barrel crankcase, dual overhead cams, four valves per cylinder, and cam drive by gear train. A highly significant new touch was the use of piston-type, or cup-type, cam followers.

Duesenberg's eight followed Bugatti practice more closely in using a single, shaft-driven overhead camshaft and three valves per cylinder, actuated by finger-type cam followers. By 1921, the Ballots had given up at Indianapolis but Duesenbergs finished second, fourth, sixth and eighth. At this point Miller decided to start building racing engines again.

Perhaps he sensed that at last his team was complete. Leo Goossen, a

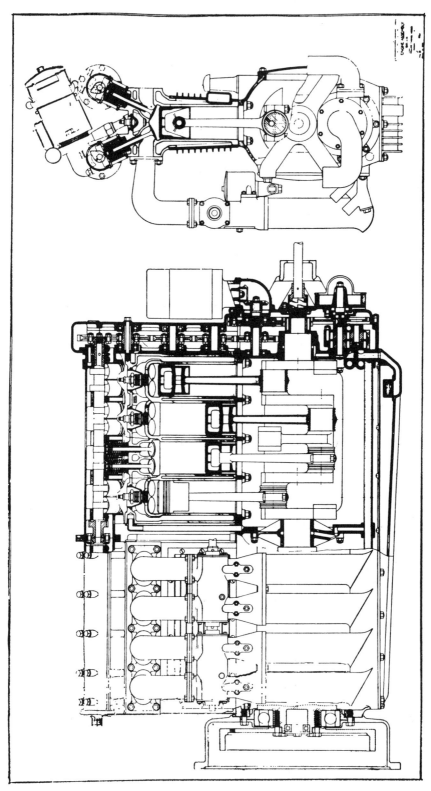

Miller's first 183cu in straight-eight engine. The camshaft was made in two pieces because of the small size of the shop's cam grinder. Note the four valves per cylinder in pent-roof combustion chambers

young draughtsman from Buick, had been with him for a year and had
proved to be pure gold. Offenhauser could fabricate anything you could
show him, but the problem was showing him what was in your head.
But with Leo, now, you could describe something to him in principle
and he would hand you back a magnificent set of drawings, complete
to the last dimension and detail, including such monotonous trivia as
gear design and stress analysis. At last there was no obstacle to the flow of
ideas from Miller's brain to finished metal.

The first Miller 'eight' was frankly derived from the Duesenberg but
contained many intended improvements. It used the Duesie's same bore/
stroke ratio as well as a three main-bearing crankshaft. But instead of the
single-cam, three-valve top-end layout it used the Ballot approach of
twin cams, four valves per cylinder and cup-type cam followers. (See
drawing opposite.) This barrel-case engine, which complied with the
then-effective 183cu in displacement formula, contained most of the basic
design features found in the present-day Offenhauser, its direct descendant. One of the first of these Miller 'eights' was installed in Jimmy
Murphy's Duesenberg car and it won the Indianapolis 500 its first
season out. Miller was launched in the Big Time.

The displacement limit for engines changed in 1923 and eleven out
of the thirty-five starters at Indianapolis that year were complete Miller
cars. Their new 122cu in engines were closely similar to the 183s, with
two big differences. First, the overly flexible three-bearing crankshaft
was replaced by one with five mains. Second, Colonel Hall, of the Hall-
Scott Motor Co, had suggested to driver Tommy Milton that the
hemispherical combustion chamber which was working out in aircraft
might be equally good in a racing-car engine. Milton passed the idea
on to Miller and it was applied. This was probably the first time in
history that the two-valve, dome-shaped (as opposed to four-valve,
prism-shaped) chamber was exploited efficiently in a racing power plant.
It worked well enough that first year for Millers to sweep the Indianapolis bricks to a first, second, third, fourth, sixth and seventh win, a fairly
devastating victory.

Fred Duesenberg, who would say disgustedly to drivers, 'So you're
going out West to get a car from those cowboys,' had an historic trick
up his sleeve for the following year. Dr Sanford G. Moss, of the General
Electric Corporation, had developed a centrifugal supercharger which
he thought would be applicable to racing-car engines and he made copies
available to Fred, along with his own expert knowledge. All this was
kept secret, and with the advantages of sudden surprise and greatly aug-
mented power, a blown Duesie walked away with the 1924 '500'. But
Miller rallied instantly. The following morning he was beating down

the doors of Allison Engineering, with Goossen at his side. They begged all the information they could get pertaining to superchargers of this type. They rushed back to Los Angeles to cut and try until they had a blower that would blow off the Duesenbergs. Experts were to say later that no one ever got more out of a supercharger than Harry Miller, who became equally the master of the Roots type, for marine use.

That year of 1924 was a busy, historic one. Jimmy Murphy had come to Miller with the order for a front-drive car. Murphy reasoned that faster lap times could be made through faster cornering, and that a car that could pull instead of push itself through the corners would get the job done best. Miller and Goossen went to work on the problem. The only precedent was the old Christie front drive, a brutal juggernaut with huge engine mounted transversely between the wheels, which were bolted directly to the ends of the crankshaft! They tried to draw up a more sophisticated version of this but Murphy almost abandoned the project when he saw it.

Miller and Goossen granted that he was right, forgot about precedent and went back to the drawing-board. This time the result was as advanced as the Christie approach was archaic. They reversed the engine in the frame, flywheel end forward. Ahead of it was Miller's positively locking, multiple-disc clutch. Then a drive shaft a few inches long, terminating in a pinion gear within what they called the front drive unit. This remarkably compact aluminium housing contained the ring gear, the differential and a three-speed-and-reverse transmission, all this being accomplished partly by means of three hollow shafts, one telescoped inside the other. Outboard of this assembly were the inboard brakes, then the inner universal joints, then the stub axles, then the outboard U-joints, steering knuckles and splined drives to the front wheel hubs. This whole articulated mass was held together by a De Dion tube which was suspended by a pair of semi-elliptic springs on each side of the chassis. The Miller front drive was a true masterpiece of engineering. It included the world's first use of De Dion suspension on a racing car. It was the first real forward step in American racing-chassis design, the first time that racing-car suspension had not been derived directly from American passenger-car practice.

Jimmy Murphy was killed in a race at Syracuse, NY, before the car was completed, so it was left to others to demonstrate the value of the principle that he had had faith in.

The following year, 1925, was a good one for Miller. Although the Indianapolis victory went to de Paolo's Duesenberg, Dave Lewis and Benny Hill finished in second place with an average of 100·82mph—just three-tenths of an mph slower than the winner. The amazing thing was that they did it in this front-drive Miller, in its first appearance at the Speedway. Seventeen Miller cars and four Duesenbergs started in the race, every one was supercharged and, in spite of record speeds, there was not a single case of blower trouble. Six Millers were in the top ten finishers and, by almost winning the race, front drive proved itself beyond all doubt. Immediate predictions made by the knowledgeable were that front drive and supercharging soon would be adopted by passenger cars and that racing cars with four-wheel-drive would appear in the near future.

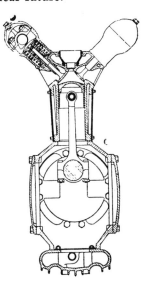

Cross-section of the Miller '91'. Note dry sump, barrel crankcase and two valves per cylinder in hemispherical combustion chambers. Bugatti adopted the Miller top end layout, starting with his Type 50

We have been speaking of large numbers of Miller entries and small numbers of Duesenbergs. This suggests that Miller's victories may have been due to his packing starting-fields with his own products, but this was not at all the case. Instead, it is evidence of further Miller pioneering. Duesenberg was essentially a private racing organisation which produced very few cars and sold hardly any. Miller, in contrast, was in the business of building thoroughbred racing cars for sale on the open market. In this respect he resembled Bugatti but, of course, on a smaller scale. Like any other series manufacturer, he published catalogues and fixed price lists for his products, which were available to anyone with the necessary money. The '91' engine listed for $5,000, the rear-drive car for $10,000 and the front-drive car for $15,000. These prices may seem high but they were actually only a fraction of the cost of earlier— and later—racing cars and engines that were produced on a one-off or few-off basis. In spite of hand techniques and the highest standards of quality, Miller was able to achieve volume and economy by standardising every part of the machine. Even the smallest bracket was fabricated according to precise engineering drawings, and interchangeability of parts compared favourably with the passenger-car field.

The following year, 1926, introduced the four-year period which was to climax and conclude what often is called the Golden Age of automobile racing in America. The formula called for a limit of 91·5cu in (1,500cc) for supercharged engines, and for it Miller scaled down his larger cars and their 122 cubic inch (2 litre), hemispherical combustion chamber engines. The Miller 91 was the highest refinement of the American straight-eight racing engine. It was an absolute jewel in design, workmanship and performance, producing a reliable and easy 250hp on methanol fuel. Bore and stroke were 2·1875 × 3in and the power peak occurred at about 8,000rpm or 4,000 feet per minute piston speed. At this crankshaft speed the screaming supercharger impeller turned at 40,000rpm. Pistons weighed 4·5oz, connecting rod and piston 19oz, crankshaft 46lb, complete engine 330lb, complete car 1,400lb. (See drawing, p 217.)

These cars dominated oval track racing almost totally and the records they established were legion. Some outstanding examples were:

Leon Duray, Indianapolis—124mph; the Speedway's most enduring lap record;
Leon Duray, Packard Proving Ground (2½ miles)—147mph (world's record for a closed course);
Frank Lockhart, Muroc Dry Lake—171mph (on pump petrol).

Lockhart mounted two Miller 91 blocks on a common crankcase in his Stutz Black Hawk and at Daytona Beach was clocked at an official 198mph. He was killed shortly afterwards in the same car when a tyre burst at about 220mph.

Lockhart's first ride at Indianapolis was in a Miller 91 in 1926. He won the race, and of the fastest ten finishers only the fifth and tenth place cars were not Millers. The next year was the last one Duesenberg was to win at the Speedway. George Souders was the chauffeur and six of the quickest ten drove Millers. In 1928 the big race was won again by a rookie. The twenty-three-year-old kid from the dirt tracks was Louie Meyer and his Miller scored the fastest average ever made by a 91 inch machine in the '500': 99·482mph. Meyer remembers that ride vividly.

'What a sweet engine it was—and it could have been developed so much further. With a bigger bore and shorter stroke, it would have had a lot more speed. The singing of the spur-cut cam drive and blower gears gave it a thrilling, hair-raising sound. With a pair of small pipes its exhaust note was the most classical of the ripping-canvas variety; with a single, four-inch pipe it had a soft, deep boom. You had to rev it up to 6,000rpm to get any performance at all, but that was exactly what we wanted. Below that engine speed, stepping on the throttle was like stepping on a sponge. The brakes were the same after you'd used them once. They looked beautiful but they were made to The Old Man's lightweight standards and would take no heat at all. After Lockhart taught us to broadslide the turns, we only used them for pit stops anyway.'

The year 1929 was a fateful one for automotive history in many ways. Leon Duray contributed his bit when he took two Miller front-drive Packard Cable Specials to Europe that summer. At Montlhéry, he racked up a few international records in the 148mph range, then moved on to the Grand Prix at Monza in September. There, without experience and with a totally unsuited car, he turned the record lap in practice, ahead of such masters as Nuvolari and Varzi and the GP Alfas, Talbots and Bugattis. The Continental motoring press freely admitted that Europe had nothing that could approach the efficiency of the American thoroughbred engine. Ettore Bugatti's son Jean witnessed Duray's performance and, on the spot, acquired both Millers from the colourful, controversial six-footer. Bugatti took the cars to Molsheim, ignored their chassis but analysed their engines with the greatest care. From that time on—from the Type 50 on—the single-cam Bugatti was seen no more. In its place was a dual overhead-camshaft power plant that faithfully copied the Miller top end. Thus the circle was completed and Miller's debt to Bugatti for the straight-eight concept was repaid.

Miller's influence extended to the passenger-car field during this period. For a change, he had taken the trouble to patent his solution to the problem of front-wheel drive. E. L. Cord, the guiding genius of Auburn and Duesenberg passenger-car production, decided to manufacture a front-drive passenger car under his own name. He retained Miller on this aspect of the design programme and made history with the L-29 Cord car. The Old Man equipped one of the first Cords with a four-cam, V-16 engine and gave it to his wife, Edna.

The economic depression that spread around the world in 1929 brought thoroughbred automobile racing to a sudden end in the States. Competition formulae that placed rigid limits on design for the sake of the improvement of the breed were replaced by so-called free formulae. The 'Junk Formula' that replaced 91 inch competiton in the USA permitted 366 cubic inches displacement, no more than two valves per cylinder, a minimum weight of 7·5lb for each cubic inch, a minimum car weight of 1,750lb, no superchargers and not more than two carburettors.

The motives behind this reversal of the whole historic trend were aimed at the stimulation of the sport by (1) making it a cheaper game to play, (2) severely handicapping high-performance machinery and making it easy for cars based on production-car engines to compete and (3) attempting to draw the American automotive industry and all its wealth back into racing. The industry responded well, but the men who knew racing best remained faithful to pure, race-bred design. More than half of the Indianapolis entries in 1930 were Miller eights, bored and stroked to give displacements ranging from 138 to 151·5 cubic inches, the limit for twin-cam engines. Of the thirty-eight starters, fourteen were powered by Miller engines. Billy Arnold won in an unblown Miller-Hartz front-drive, averaging 100·448mph. Louie Meyer came in fourth in the Sampson Special; its power plant was the Lockhart 'twin 91'. Second place was taken by Shorty Cantlon in a car powered by a novel engine, a Miller four. This was the first faint rumble of the Offy era to come.

Back in 1926, Miller had designed a marine engine at the request of the boat-racing champion Dick Loynes. He needed a four-cylinder engine to beat the modified Fords and Stars which had been developed to their absolute limits. Miller came up with a 'four-banger' of the required 151 cubic inches, using two valves per cylinder in hemispherical combustion chambers. The engine was much heavier and much more robust than typical Millers, since Loynes had specified a power plant that could be developed to higher outputs over the years. This engine made champions in small-displacement inboard racing and, by adding four-cylinder blocks to suitable crankcases, Miller was able to offer eight- and sixteen-cylinder marine racing engines with displacements of 310 and 620 cubic inches.

As unblown Miller eights continued to dominate the board and dirt tracks under the Free Formula, Bill White, one of the key figures in American racing history, conceived the idea that the Miller four had greater potential high-speed torque. Miller was in the East but White approached Goossen and Offenhauser and asked for help in modifying a Miller marine four for track racing. The help was gladly given, the engine's displacement was increased to 183 cubic inches and its bugs were worked out during months of exhaustive testing at Los Angeles' Ascot Speedway, with Cantlon at the wheel. White's faith was rewarded at the Speedway in 1930. He was jeered at for running a 'four-banger' against a field of sixes, eights and sixteens. But the four's inexorable torque kept it comfortably in second place for nearly every one of the 200 laps.

Miller had sold his business to a short-lived firm in 1929 and had gone

East to seek a new fortune. He did not find it and returned to Los Angeles in 1931, disgruntled with the liberties that Goossen and Offenhauser had taken with the marine four on White's behalf. Still, he founded the Rellimah Company (H. A. Miller spelt backwards) around the old Team and entered the twilight of his career.

One of the last creations of the Team was one of the most suddenly successful. In March of 1932, Harry Hartz roared into the shop with a frantic order for a 183 cubic inch straight-eight to be run at Indianapolis two months later. There definitely was not even time to run the engine before it was crated and shipped East. At the Speedway, it was dropped into Hartz's car, qualified, won the race and remained one of the foremost performers for years to come. This engine, essentially a scaled-up '91', was cut in half to make the Offy 'Midget'. Goossen made the first proper drawings for it in the summer of 1933 and the first 97 cubic inch 'four-banger' was made in 1934. Miller went bankrupt in 1933 but before he faded from the West Coast scene he blazed out in glory with a V-16, two four-wheel drive cars and several with De Dion rear ends for the Speedway. He also found time to design and build a juggernaut of a 1,113 cubic inch V-16 marine racing engine for Gar Wood.

Miller went straight to Indianapolis to make a new start. One of his first contacts was with Preston Tucker, later the driving force behind the imaginative but ill-fated Tucker Torpedo car. Tucker and Miller hatched a plan to build front-drive racing cars around the then-new Ford V-8 engine. When Tucker finally sold the Ford Motor Company on the project it was February of 1935; the deal called for ten cars—all to be built in less than three months. The few that were completed in time for the race were as beautiful as they were technically exciting. They were called the best-looking, best-streamlined cars ever seen at the Speedway and were particularly remarkable for independent suspension of all wheels. That a few cars were functional in time for qualification trials was a heroic accomplishment in itself but, of course, there was no time to begin to get them sorted out. Ted Horn's mount was lucky to finish sixteenth. The project had been authorised by Edsel Ford without his father's knowledge. Old Henry decreed its immediate termination.

This blow left Miller discouraged, depressed and temporarily beaten. Meanwhile, Fred Offenhauser bought some of the Rellimah machine tools and the drawings and patterns for the Bill White-type 'four-banger' and began eking out a threadbare existence building engines and parts for the racing fraternity, with Goossen's consulting services always on tap. In 1937, Miller returned to the fray and designed a pair of all-independent suspension cars around two 255 cubic inch fours which he bought, ironically, from Offenhauser. The cars bristled with

N

novel features: side-mounted fuel tanks, cartridge starting and externally mounted radiator cores were among them. But Miller lost interest in these machines before they were finished and they were too slow to qualify for the big race. His attention was diverted by the Gulf Research & Development Company's decision to build and field a team of cars that would demonstrate the excellence of Gulf's oil and pump petrol. Miller landed the designing job and at last had a blank cheque for his boldest dreams.

The Gulf-Millers may have borrowed their chassis layout from Germany's Auto Union but no more. For them, The Old Man designed a 180 cubic inch, six-cylinder, rear-mounted engine with an over-square bore/stroke ratio. The block was inclined at an angle of 45 degrees to reduce the car's centre of gravity. A single light-alloy casting embraced cylinder block, head, crankcase and camshaft housings. Thin-wall cylinder sleeves were shrunk into the aluminium bores. A large centrifugal supercharger with inlets at both front and rear drew through two carburettors and pumped the fuel through a large intercooler to the inlet ports. Output on 81 octane fuel was 246bhp at 6,500rpm. A four-speed transmission, plus a three-ratio final-drive selector, fed power to both the front and rear wheels. The fuel tanks were built into the deep frame channels and the brakes were a pioneer disc type of original Miller design.

Four Gulf cars were built and raced but were jinxed. Three were entered in the '500' in 1939. One was wrecked and burned. Another spun out and the third, the only actual qualifier, ran forty-seven laps and then dropped out because of a broken valve spring retainer, it is said. The next year, two cars were entered. One was wrecked and burned and the other was withdrawn because officials decided that the fuel-tank location created a fire hazard. In 1941, two cars qualified but one was lost in a garage fire and the other retired with a locked gear-shifting mechanism. The best that could be claimed for these magnificent machines were the dozen international records which George Barringer established with one in 1940, at Bonneville. His best speed was 158·446mph for the flying five kilometres—on Gulf No-Nox gas.

And so Miller moved on to Detroit, looking for another chance. He was full of ideas, all radical, all years ahead. Three years previously his comments on the passenger-car industry's offerings had appeared in the automotive press:

> The entire principles ought to be changed. The engine should be in the rear to eliminate the long drive-shaft, torque tubes and miscellaneous rods under the body. It should be radically streamlined, faster, hydraulically braked and lighter. We are definitely headed for a lighter car that will start

and stop more readily. It takes considerable horsepower to start a heavy car. Lighter ones will be more economical and safer.

A few years after Miller's death, Eddie Offutt described to me in detail the projects that he had helped The Old Man with during the last couple of years of his life.

'One was a fantastic, infinitely-variable blower drive for aircraft; he had completed the engineering drawings and calculations. The other was a small, front-drive, light car with high performance characteristics. For years Harry had said that sports cars were the coming thing. The car he had designed was to weigh between 1,600 and 1,800lb. Its double overhead camshaft engine would put out between 90 and 100hp. The engine sat transversely, and was inclined to the rear about 45 degrees. All weight was sprung, except for the short stub shafts. His idea was to put the weight where it would be needed for acceleration, so that it would shift back on the axles. He planned to use a fluid drive along with an automatic transmission of his own design. He had much of it built in prototype form, and I think that his automatic transmission had all the promise that present-day automatics have. He planned to put this under an aerodynamic coupé body on a wheelbase of about 112 inches. It was a very low, cute car and he meant it for large-scale production.'

That was almost thirty years ago and Miller was already anticipating the Mini-type vehicle, along with variations that he thought the public would embrace. He was still pioneering, as he had pioneered with motorcycles, outboard engines, sparking-plugs, carburettors, fuel pumps, aero-engines, automotive and marine engines of every configuration, light alloys, metallurgy in general, hydraulic brakes, metallic brake linings, disc brakes, supercharging, front-wheel drive, De Dion suspension, full independent suspension, four-wheel drive, variable final drive and inclined engines—to name a few of his fields of *avant garde* activity.

Arch-conservative Offenhauser, who thought in terms of developing a single engine design over a period of years or even decades, used to say of Miller, 'Harry always liked to shoot the works on something tricky.'

Offutt put it differently. 'The man could see so much farther into the future than any of us. He had the courage and capacity to dream in the grandest manner. We all dream, but few of us do so with any hope of our dreams becoming reality. Harry carried countless dreams straight through to realisation in beautiful, living metal. By working with him, we who shared his dreams realised them too.'

8 VITTORIO JANO

by PETER HULL
and ANGELA CHERRETT

VITTORIO JANO (1891–1965). Italian: Sometime
designer for Fiat, Alfa Romeo and Lancia.
Outstanding creations include Alfa Romeo P2
and P3 Monoposto, and Lancia D50 GP cars; also
Alfa Romeo 1500, 1750, 2300 and 2900, and
Lancia Aurelia sports cars.

VITTORIO JANO

FOR ECONOMIC REASONS, Italy was a late starter in the motor industry although, significantly, motor races and competitions were being held on Italian roads before the turn of the century.

Turin soon became the centre of motor manufacturing, largely due to the activities of two remarkable brothers called Ceirano, Giovanni and Matteo. In the nineties, they were established as bicycle manufacturers, and they also built a few belt-driven cyclecars. In 1899, their small Turin factory, employing about fifty men, was taken over by a syndicate led by Giovanni Agnelli, and formed the nucleus of the great Fiat empire. Two young men whom Fiat inherited from the little factory both later became famous as racing drivers and car manufacturers—Vincenzo Lancia and Felice Nazzaro.

The Ceirano brothers did not stay long with Fiat, but left to form what amounted to a series of car factories in the Turin area. In 1904, Matteo started the Itala concern which, by the time of World War I, was second only to Fiat in influence. By 1906 Matteo had moved on to form SPA, and Giovanni founded STAR (Societa Torinese di Automobili Rapid), makers of a very good car called the Rapid, and then he made the SCAT car, which later became known as the Ceirano. Ironically, several of the makes founded by the Ceiranos were eventually absorbed by Fiat.

In the field of car design, Italy was not a significant contributor in the pioneering sense, both Fiat and Itala being content to model themselves on the successful Mercedes in the early years of the century, and it was not until 1903 that Fiat ventured abroad to represent Italy in international motor racing. The year 1907 was the first great year for Italy in motor racing, when the Fiat driver, Felice Nazzaro, won all the most important races: the French Grand Prix, the Kaiserpreis and the Targa Florio. By this time Fiat had broken away from the Mercedes tradition in the design of their racing cars, and their engines featured push-rod operated overhead valves instead of the Mercedes inlet-over-exhaust system. More successes were later attained by huge Fiat racing cars with single overhead camshaft engines, but these 'last of the giant racers' were outdated from 1912 by Ernest Henry's smaller and more sophisticated Peugeots. Carlo Cavalli was in charge of Fiat design at this period, but Ettore Bugatti is thought to have had a hand in the Fiat engine drawings.

Bugatti was a Milanese and a designer of great significance, but most of his work was done abroad, and did not directly benefit the industry of

his native land. In the years up to 1914, Giulio Cesare Cappa and Gius-
tino Cattaneo were two of Italy's most forward-looking designers.
Cappa was designer for the Turin firm of Aquila-Italiana, before going
to Itala. His six-cylinder Aquila-Italiana of 1906 was most advanced,
with a ball-bearing crankshaft, overhead inlet valves, monobloc engine
without a separate gearbox and, most remarkable of all, aluminium
pistons.

Cattaneo's finest car was a sixteen-valve single overhead camshaft
sporting Isotta-Fraschini, which first appeared in 1910, Isotta, incidentally, being the other Italian make besides Fiat with which Bugatti's
name was associated. By 1914, all Isotta models had front wheel brakes,
the firm having first introduced them as early as 1909.

It was in the twenties, however, that Italian design really blossomed,
particularly in the racing-car field. Cappa was one of those who had a
hand in this, but the most famous name amongst all the Italian designers who worked in the first half of the twentieth century is almost
certainly that of Vittorio Jano. And it was in 1909 that the eighteen-
year-old Jano obtained his first job at Giovanni Ceirano's Rapid works
in the Via Nizza in his native Turin.

Vittorio Jano was born in 1891 in the town of San Giorgio Canavese,
near Turin, his birthday being on 22 April. His name is believed to be
Hungarian in origin, originally Janos, the family having moved to the
Turin area around 1760. They were a military family, and a combination of both military and technical interests is appropriately shown by
the fact that Vittorio's father was in charge of one of Turin's arsenals.

Inevitably young Vittorio was technically minded, and before joining
Rapid as a draughtsman he had completed a course at the Istituto Pro-
fesionale Operaio in Turin. At that time he acquired a Minerva motorcycle, which must have first instilled in him his genuine love of driving.
It was T. E. Lawrence ('Lawrence of Arabia') who said that a motorcycle gives a better appreciation of speed to its rider than a motorcar,
speedboat or aeroplane, and the remark of the distinguished British
actor, Sir Ralph Richardson, to a newspaperman who confessed he had
never been on a motorcycle is short, but expressive: 'My dear fellow!
You haven't lived!'

By these standards, Vittorio had definitely lived when, in 1911, he
left Rapid and transferred to Fiat, where his ability was soon recognised
by the chief engineer, Guido Fornaca, who put him to work in the
design department under Carlo Cavalli.

The American writer, Griffith Borgeson, to whom we are indebted
for detail of Jano's early life and origins, has told how Cavalli recognised
in the young Jano a rare determination and drive, a way of reflecting

Page 229 (*above*) Vittorio Jano in the mid-twenties when he was first earning fame as designer of the P2 Alfa Romeo Grand Prix car; (*below*) the Fiat Model 501S, a sports version of the 1,460cc side-valve four-cylinder touring car. Jano was associated with its design under Carlo Cavalli

Page 230 (above) P2 GP Alfa Romeo at the 1924 Italian Grand Prix. Jano is on the far left and the man with the large moustache standing behind the radiator is Nicola Romeo; *(left)* the 6C 1750 Gran Sport Alfa Romeo in process of winning the 1930 Mille Miglia with Tazio Nuvolari at the wheel and Battista Guidotti beside him; *(below)* Varzi makes a pit stop in the 1934 Italian GP at Monza driving a GP Monoposto Tipo B Alfa Romeo (P3)

Page 231 (above) Jano's 12C 36 racing car was similar in appearance to the earlier 8C 35 GP car but had two exhaust pipes underneath instead of the single one on the offside which distinguished the eight-cylinder car; (below) Jano's last GP design for Alfa Romeo, the GP Tipo 12C of 1937, was never fully developed. A rear-axle failure in the 1937 Italian GP presaged Jano's departure from Alfa Romeo

Page 232 (above) Piero Taruffi airborne in the D24 Lancia sports racing car during the 1954 Mille Miglia won by his team-mate Ascari in a similar car developed by Jano from his Aurelia design; (below) D50 Lancia Grand Prix cars photographed during practice for the 1955 Italian GP

profoundly on design problems and finding original solutions to them, an elegance in the conception and execution of his work, and a ruthless objectivity and capacity for self-criticism. Borgeson says that Cavalli made Jano his personal pupil, and to the end of his days Jano acknowledged his debt to his old master.

By 1917 Jano was a senior draughtsman at Fiat, but was also concerned with the organisation of their racing teams. We know that he was at Lyons for the famous French Grand Prix in 1914, where Fiat entered a team of three small, low cars with streamlined tails, looking more like the GP cars of the early twenties. They had single overhead camshaft 4½ litre 80bhp engines in unit with their gearboxes and four-wheel brakes, but still retained chain drive as Fiat reckoned that the low unsprung weight of a chain-drive rear axle gave good roadholding on racing cars. Despite all this they did not, however, emulate the success of the Fiat racing cars of the early twenties. This early evidence that Jano was recognised as a good organiser as well as a technician is important.

Fiat's most remarkable racing car was the 2 litre, six-cylinder Type 804 of 1922, of which the late Laurence Pomeroy wrote:*

> The complete superiority of these cars over contemporary 2 litre models, and the impression they made on technical opinion, can be likened to the effect of the first Mercedes in 1902, and the 1912–13 Peugeot models designed by Henry.

These revolutionary, very light, high-revving small Grand Prix cars were truly invincible, and a model for all their competitors. At this time the Fiat racing design department had a monopoly of brilliant young men, including Vittorio Jano. He was unique, however, in having exceptional drive and organising ability in addition to his technical gifts.

The main credit for the design of the successful four-, six- and eight-cylinder racing Fiats of this period must go to Giulio Cesare Cappa and his assistant Tranquillo Zerbi, working under Cavalli. It is interesting that Cappa pioneered engine/gearbox unit construction in Grand Prix cars, bearing in mind his Aquila-Italiana design which had this feature in 1906. He also pioneered supercharging in Grand Prix racing.

Jano was not only concerned with racing-car design, but also helped Cavalli with the production of the Fiat touring cars, notably the famous four-cylinder side-valve 501. Anyone examining the original drawings for the 510 and 520 six-cylinder side-valve Fiats, will find Jano's is the

---

* *The Grand Prix Car* by Laurence H. Pomeroy, Motor Racing Publications Ltd, 1949.

third signature on them, just as Luigi Fusi's was to be on many Alfa Romeo drawings.

In 1923, when Jano was in charge of Fiat's racing team, he was asked to join the Milan firm of Alfa Romeo, which was already on the ascendant in both the production and racing spheres. The Alfa Romeo firm had originated in 1909, when some Milan businessmen decided to found a firm to manufacture a 100 per cent Italian car in works at Portello which had formerly been used to build small French Darracq cars under licence. These cars had such a poor reputation that Italians who could afford to do so preferred to buy their Darracqs from the French factory at Suresnes, finding the higher cost of them was justified. Thus it was not surprising that the Italian branch of Darracq went out of business.

The new firm was called Societa Anonima Lombarda Fabbrica Automobili, and consequently their cars were known as ALFAs or Alfas. To design them, they employed thirty-seven-year-old Giuseppe Merosi, a first-class technician born in Piacenza, who had formerly worked for Fiat in Turin, and later was chief designer for Milan's only important car manufacturer, Bianchi.

Alfa started racing in 1911, the same year as Bugatti, but had no outstanding successes, though they had a good driver in Giuseppe Campari. In 1914, Merosi designed a genuine twin-overhead-camshaft Grand Prix car, but it was not raced until after the war, and then only in Italian events.

In 1915, the Alfa factory was taken over by Nicola Romeo, an industrialist from Naples with an engineering background, who had made his money by manufacturing compressed-air mining machinery and earth-moving equipment under licence from the American firm of Ingersoll-Rand. During the war, he diversified and greatly expanded his activities, making tractors, railway wagons and even aircraft and aero-engines.

He was enthusiastic about motorcars and motor racing, and as early as 1919 the pre-war Alfa cars were being entered in events under the new name of Alfa Romeo. In 1920, Campari scored the first win for Alfa Romeo at Mugello with a racing version of the pre-war, push-rod, overhead-valve 40/60hp model, and in that year Enzo Ferrari came second to Meregalli's Nazzaro in the Targa Florio, driving another 40/60hp.

By 1921, an efficient racing department had been built up at Alfa Romeo, manned by Enzo Ferrari, Antonio Ascari, Giuseppe Campari and Ugo Sivocci. A new sports version of the pre-war, side-valve 20/30hp was evolved by Merosi, at the instigation of Antonio Ascari, known as the ES Sport, and was very successful in competitions.

Merosi's best design, however, was his post-war, six-cylinder, push-rod, overhead-valve 3 litre RL, in touring, sports and racing form. It became quite a prestige car in Italy. Mussolini ran a sports version, and exotic personalities such as the Aga Khan and the Crown Prince of Siam were associated with it. In 1923, Ugo Sivocci won the Targa Florio with a special racing version known as the RL Targa Florio model, and later examples proved a match for the current 2 litre super-charged Mercedes racer and won important Italian races driven by Ferrari, Ascari and Count Masetti in 1923 and 1924.

This type of racing, though, was not enough for Nicola Romeo, who was anxious for Alfa Romeo to enter the international lists with a pukka Grand Prix car. An obvious way of setting about the design of a com-petitive car was to try to attract some of the brains and know-how from the Fiat factory at Turin—an idea that had also occurred to Louis Coatalen of the English Sunbeam firm. Coatalen had engaged Bertarione of Fiat to design the 1923 Grand Prix Sunbeams, one of which won the French Grand Prix at Tours, driven by Dehane Segrave.

It so happened that, after this race at the beginning of June 1923, one of Fiat's brilliant young men, Luigi Bazzi, had had a row with Ing Fornaca, Fiat's general manager. Bazzi and Jano had been great friends at Fiat and had often shared test drives in both the touring and racing cars. Bazzi was also a close friend of Ferrari, and after the disagreement at Tours, it needed little persuasion from Ferrari to make Bazzi transfer from Fiat to Alfa Romeo in Milan.

At Portello, Bazzi immediately joined Merosi in completing the de-sign of a 2 litre, six-cylinder Grand Prix Alfa Romeo, officially known as the GPR 1923 (Gran Premio Romeo 1923), which had been on the stocks since the previous autumn. Unofficially, the title of this car was shortened to the 'P1' by the popular press.

The design of the P1 was inspired, like Bertarione's Sunbeam, by the all-conquering six-cylinder Type 804 GP Fiat of 1922. In 1923, the Type 804 was replaced by the eight-cylinder supercharged Type 805 Fiat. Thus, although the 112mph P1 had proved satisfactory on test, the team of three P1s which were entered for the 1923 European Grand Prix at Monza on 9 September were already outdated in design. The Sunbeams had been fortunate at Tours, where the Type 805 Fiats had trouble with their Wittig vane-type superchargers, whereas for Monza they were fitted with reliable Roots-type blowers which increased the brake horsepower from 130 to 145. The output of the P1 was a mere 80bhp, 12bhp less than that of the Type 804 on which it was modelled.

Tragedy overtook the P1 team before the race at Monza, for Ugo Sivocci left the course on a turn during practice and was killed. As a mark of respect, the whole team was withdrawn from the race, and thereafter the P1 never appeared before the public. In the European GP, as expected, the Type 805 Fiats proved victorious.

Sivocci's unhappy death did not deter Nicola Romeo in his ambition to field a Grand Prix team, and for 1924 he was left with the choice of developing the six-cylinder P1 and supercharging it, as Sunbeams were doing with their 1923 car, or else building a new eight-cylinder super-charged car on the lines of the Type 805 Fiat.

It was Luigi Bazzi who suggested that the one man who was capable of the task of organising the design and building of a completely new car in the short time available was Vittorio Jano, *if* he could be persuaded to leave Fiat.

Ferrari had never met Jano, and he has told in his memoirs how he called at Jano's house in the Via San Massima, Turin, one day in September 1923. Jano was out, but his wife Rosina told Ferrari that she was sure that Jano was too much of a Piedmontese ever to think of leaving Turin. Soon Jano himself arrived, and Ferrari explained to him the re-sponsible task Alfa Romeo had in mind for him, at a greatly increased salary over the one he was receiving from Fiat.

It was not the extra money that attracted Jano, but the chance he was being offered to allow both his technical and organisational gifts to be given a free rein. At Fiat, he had his own ideas for improving their racing cars, but Fiat were content to leave them as they were. However, he was not willing to negotiate with Ferrari and told him 'Unless Romeo himself comes to see me, I'm not going.'

Romeo did not come to see him himself, however, but sent his per-sonal assistant, Ing Giorgio Rimini to Turin the next day, accompanied by Ferrari, and the necessary papers were signed. In a matter of days, Jano, with Rosina and their two-year-old son, Francesco, had moved into an apartment on the first floor of No 71 Corso Sempione in Milan.

When Jano first entered the technical office at Alfa Romeo with Merosi, all the staff wondered who he was. He walked through the office and picked out the youthful Luigi Fusi and one other member of the staff. Fusi, who had joined Alfa Romeo in 1920, started work with him on 20 September 1923. Other technical staff from Fiat of less renown than Jano came to work at Alfa Romeo, and on 1 December 1923, Fusi was joined by Bernacchi, Graziosi, Molino and Pedrazzini. In all, the P2 team consisted of ten design and drawing staff.

Shortly after Fusi was transferred to Jano's department, Jano asked him if he liked drawing. Fusi replied that he did, and was asked if he could come to work the next day, which was a Sunday. Fusi said he would and, after attending 6 am Mass, he arrived outside Jano's house

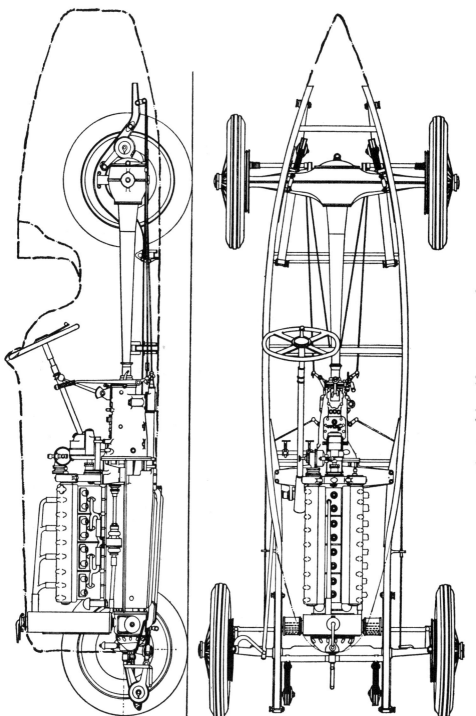

Chassis of the Alfa Romeo P2 of 1924

at 8 am. Together they went to the office, where Fusi was positioned at the drawing-board and Jano stood behind him at a table. Jano then proceeded to give Fusi the dimensions for the transverse section of the P2 engine. (See drawing, p 240.) By lunch-time they had half of it drawn out. This was on about 1 October 1923, and by 10 October much of the engine design and drawing had been done. They never made a complete drawing of the chassis, etc, but once in the metal everything went together perfectly. Fusi says that Jano was an excellent co-ordinator when producing a new car, so that the right pieces were all ready on the assembly line at the right time.

The first P2 engine was running on the bench at the end of March 1924, and the first complete car was finished by the end of May. On 9 June Antonio Ascari, with Luigi Bazzi as his riding mechanic, won the 200 mile Circuit of Cremona race on a P2 with ease from a Chiribiri and a Bugatti at 98·31mph, and was timed at over 121mph over the 10 kilometre straight.

The second appearance of a P2 was less fortunate, for Campari retired in the Coppa Acerbo at Pescara on 13 July, but the day was saved for Alfa Romeo by Enzo Ferrari, who won from a supercharged Mercedes and an SPA driving one of Merosi's RL Targa Florio cars.

The first real test of the P2s was to be on 3 August, the occasion of the European Grand Prix at Lyons, over much the same circuit as the historic 1914 French GP which Jano had attended with the Fiat team. The 1924 race proved to be no less historic than the 1914 one, as it contained one of the finest and most varied entry lists ever obtained for a Grand Prix.

Jano approached the race with some thoroughness. He had already used one of the old 1923 P1 cars as a sort of mobile experimental test bed for the P2, and by supercharging the engine had increased the output from 80bhp to 118bhp, and the maximum speed to over 120mph. A month before the Grand Prix was due to be held at Lyons, a P1 was used for the practice period so that Jano could decide which gear and axle ratios would be best suited to the circuit. On race day, all the Alfa Romeo team, headed by Ing Giorgio Rimini and Jano, were busy in the pits at first light. The result of that race is now history, Campari becoming the first man ever to win a Grand Prix on a make of car that was entirely new to Grand Prix racing, beating powerful teams from Sunbeam, Fiat, Delage and Bugatti.

This victory caused some consternation at Fiat, as all their Type 805 cars retired except one, and that finished in eleventh place. They even sent the police round to the technical office at Alfa Romeo to look for stolen Fiat drawings but, needless to say, the search was fruitless. After

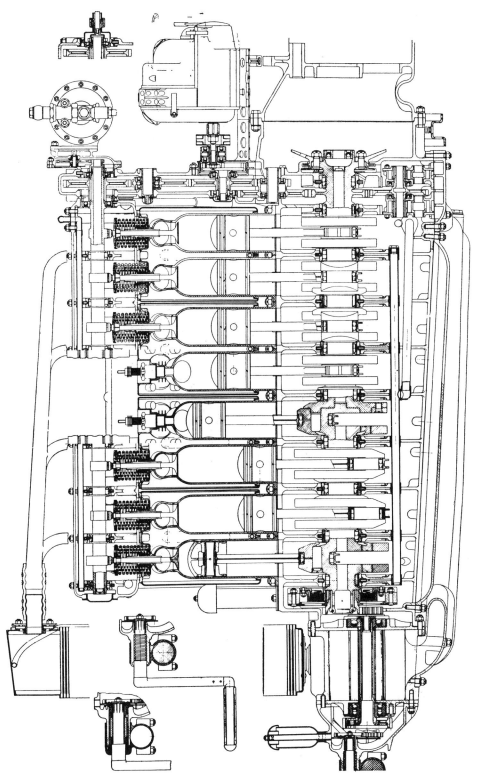

Longitudinal section of the Alfa Romeo P2 Grand Prix engine

this, Fiat evaluated their position in Grand Prix racing and decided to withdraw, so that the Grand Prix crown passed from Turin to Milan.

Both the P1 and the P2 bore distinct similarities to the Fiats, but were not carbon copies. Perhaps the most conclusive proof of this was the inordinate pride with which Jano always regarded the P2, which was the favourite of all his designs. Some of his later cars were more unique to Jano than the P2, and were even more successful, so had the P2 been a slavish copy it is unlikely it would have held such a special place in his heart.

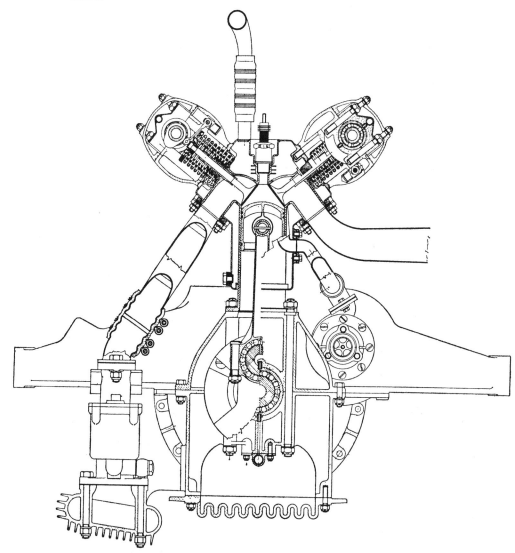

Cross-section of the Alfa Romeo P2 Grand Prix engine

In 1924 a P2 won the GP of Italy at Monza, and in 1925 the GP d'Europe at Spa and then the GP of Italy for the second time running. As a result of this last race, Alfa Romeo were declared World Champion manufacturers.

Two incidents in connection with the P2 show facets of Jano's character. The first was his edict that the P2's first appearance was to be a great surprise. This was meticulously obeyed, and Giovanni Canestrini has told us that he was as surprised as anyone when he saw the car at Cremona, although he was very friendly with Antonio Ascari at that time, and had been in his company a great deal. He knew that Ascari, together with Campari, Marinoni and Ramponi, were testing a new racing car, but he never expected to see the finished article at Cremona. He says it was not really finished even then, the bodywork looking crude and unpainted.

The second incident is the well-known one which took place during the 1925 GP d'Europe at Spa, when the P2s were having a walkover victory after all the V-12 Delages had retired, and the Belgian crowd were jeering at the Italian team. In reply, Jano immediately called all his cars into the pits, made the drivers sit round a table to eat a meal whilst their cars were cleaned and polished, and then sent them off to finish and win the race.

Jano's almost Prussian discipline, perhaps inherited from his military ancestors, has been mentioned as being instrumental in getting the P2 project through on time. Despite this, he seems to have always engendered respect in those who worked for him. It was not easy to get to know him well as he tended to be introspective and withdrawn in manner, and liked to spend as much time as possible with his wife and small son. But those who got to know him really well, like Luigi Fusi, felt a great warmth towards him.

In 1925 some details of Jano's first touring-car design for Alfa Romeo were revealed, the 1½ litre single-overhead-camshaft 1500 model, from which were derived the twin-cam 1500 and 1750 models. In April, 1926, he replaced Giuseppe Merosi as Alfa Romeo's chief designer, and Merosi went to Mathis, in France. Merosi was a sound technician whereas Jano was a brilliant one, and it must have been hard for the older man to leave the firm for which he had done so much right from its inception over a period of sixteen years. Jano was destined to face a somewhat similar situation when Wilfredo Ricart replaced him at Alfa Romeo some ten years later. Merosi was undoubtedly responsible for the solid basis upon which Jano was able to build, and much of the Alfa Romeo tradition was due to him.

During the latter half of 1926, Jano co-operated with Ing Somazzi of

o

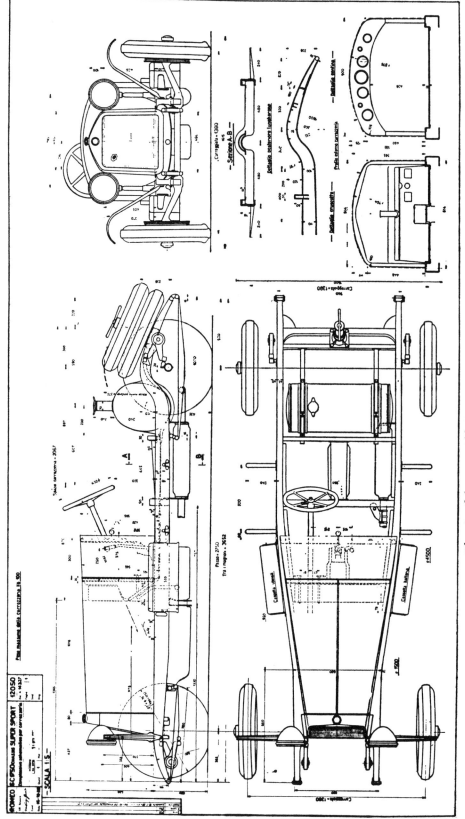

Chassis of the 1920 Alfa Romeo 6C 1750 Super Sport

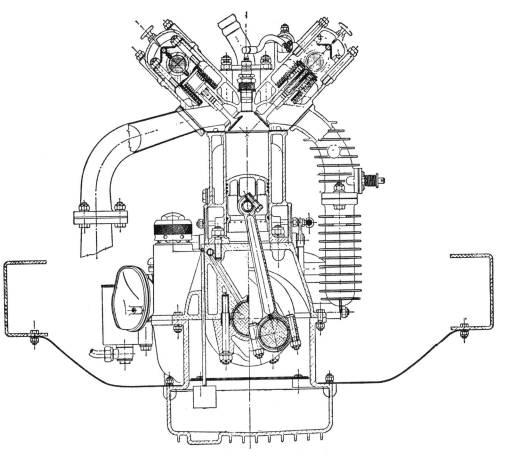

Cross-section of the Alfa Romeo 6C 1750 Gran Sport engine

NAFTA, the Italian branch of Shell, in providing a superior fuel to the benzol/alcohol mixture then in current vogue, and the experiments were very successful, finally leading to the 95/100 octane Shell Dynamin fuels, first used in the 1932 Mille Miglia.

Although we are only concerned with his work as a car designer, it should not be forgotten that, when he was with Alfa Romeo, Jano, as chief designer, was also responsible for the firm's aero-engine and commercial vehicle activities.

The exquisite 1500 and 1750 models in twin ohc supercharged form were derivatives of the P2, though they had six cylinders instead of eight, and plain instead of roller-bearing crankshafts. The big Bentley and Mercedes sports cars, which were so successful in racing when these sports Alfas first appeared, had more in common with the cars in the 1914 Grand Prix at Lyons, whereas the smaller, lighter Alfas were a sports-car interpretation of the racing cars which ran there a decade later.

The new Alfa Romeos were extraordinarily successful in sports-car races, winning both the Mille Miglia and the 24 Hours Race at Spa three years running, 1928–30, against all comers, and all the major British sports-car races such as the Six Hours and Double-Twelve Hours races at Brooklands, and the TT and Phoenix Park races in Ireland on handicap.

So successful were they that some writers have implied they were faster on these circuits than the Bentleys and Mercedes. This is not true, as they were too much down on capacity by comparison, and experienced drivers in the bigger cars could put in faster laps. The Alfa Romeo teams held two trump cards: the extraordinary skill of their Italian drivers, and the reliability of the cars. It was reliability that beat the faster 2·3 litre supercharged Bugattis at Spa, as well as the much larger Minervas. As for the drivers, more road races were held in Italy than in any other country during the twenties, and many of the Alfa drivers were used to handling really fast Grand Prix cars, capable of 140mph, under racing conditions, on ordinary roads, which gave them a great advantage over nearly all other sports car drivers.

A P2 won its last big race in 1930, the Targa Florio, with Varzi driving. Jano had modified the chassis of three of the original six GP cars for this year utilising 1750 parts, and had increased the power over the years from 140bhp to 175bhp.

In the spring of 1930 Jano began the design work on his celebrated supercharged straight-eight cylinder 8C 2300 Alfa Romeo, which first appeared in 1931, when Jano was forty years of age. In the autumn of 1930 he was engaged on designing his extraordinary Type A Monoposto racing car, which was a 'twin-six' cylinder, having two 1750 engines side by side and geared together, with two gearboxes and two propeller shafts. In some designers' hands this might have been a gormless monster, but Jano produced a car capable of beating Chiron's 2·3 Bugatti in a road race, the 1931 Coppa Acerbo at Pescara, Campari being the driver of the Type A.

The 'Monza' was the Grand Prix version of the 8C 2300 sports car, and the P3 Type B Monoposto was derived from both the Monza and the Type A. An early description of the Monza in a British magazine stated that its engine had a cast-iron block, and this has often been repeated since. In fact, only the prototype 8C 2300 engine had a cast-iron head and block, all subsequent eight-cylinder sports and racing cars having these components in alloy. Experiments were made with alloy heads on the Type A, and this experience was made use of in the 8C 2300.

Canestrini has emphasised that Jano was very even-tempered, and never allowed himself to get annoyed over anything, though this resolve was severely tested at the 1931 Belgian GP at Spa where his new 2·3 litre, eight-cylinder Monza cars were matched against the Type 51 Bugattis. One of the Alfas stopped during practice and it was found that the connecting rod of No 1 cylinder was broken. The rod was replaced and after about five laps in the subsequent practice session No 1 rod broke again, in exactly the same place, near the centre of the rod. Jano thought this indicated a batch of faulty rods and sent a telegram to Milan: 'Sack all technicians responsible.' Nevertheless, he was still puzzled about it when talking to Canestrini later in the day. Canestrini suggested that the cylinder bore might be out of line, but Jano dismissed the idea. However, two days later he told Canestrini that this was, in fact, what had happened, as when the engine was fitted with another block no further trouble was experienced. Canestrini adds: 'Presumably Jano sent another telegram to Milan . . .'.

The 8C 2300 was practically unbeatable in sports car races in the early thirties, winning Le Mans from 1931 to 1934 inclusive and the Mille Miglia in 1932 and 1933. It proved faster on a circuit than the old Bentleys and Mercedes with their much bigger engines had been but, unlike the 1500 and 1750, it was unable to beat the handicappers in British races, where the little MGs reigned supreme. Jano himself accompanied the Alfa Romeo team to Belfast for the 1931 TT, when Borzacchini lost to an MG by five seconds after breaking the lap record for the course.

From 1930 to 1937 Enzo Ferrari's Scuderia was given the responsibility of running the Alfa Romeo racing cars. Jano was extremely upset at this handing over of his cars to an outside authority (albeit one closely connected with Portello) because, although he continued to be much in evidence wherever the Scuderia was racing, he would have preferred the racing team to be solely under his own control.

Although the 8C 2300 sports cars were virtually invincible in sports car races, the Monza GP cars did not have such a clear-cut advantage over their rivals, and were beaten on occasions by both Maseratis and Bugattis. However, Jano's P3 Monoposto Grand Prix design put Alfa Romeo completely on top in GP as well as sports-car racing in 1932. The split propeller shaft design of the P3, with the differential being behind the gearbox and thus part of the sprung weight, made for a very light rear axle. One wonders if Jano was thinking back to the 1914 GP Fiats when designing this, on which low unsprung weight at the back end had been achieved by the retention of chain drive.

The P3 engine in 2·6, 2·9 and up to 3·8 litre form was similar to that

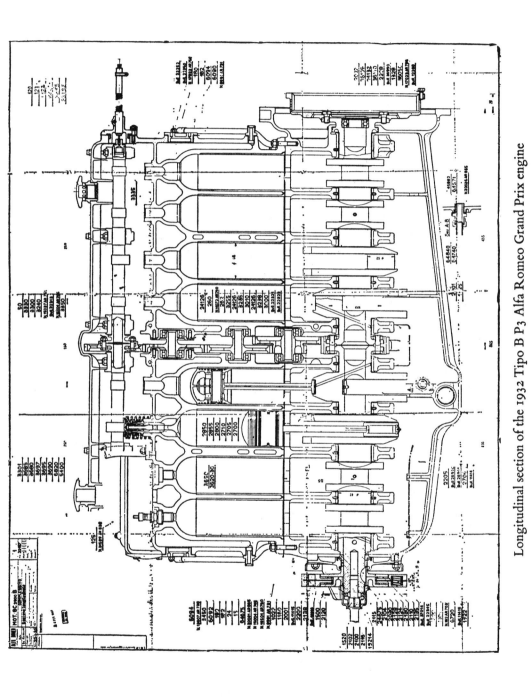

Longitudinal section of the 1932 Tipo B P3 Alfa Romeo Grand Prix engine

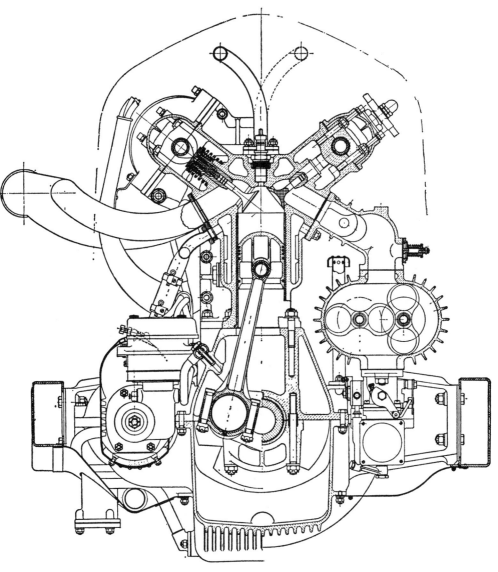

Cross-section of the 1932 Alfa Romeo P3 engine

fitted to the Monza except that it had a fixed head and twin super-chargers, but it was also utterly reliable due to the fact that no attempt was made to extract a high-power output from it. If the P3 had a weakness it was in the gearbox, which in its original form was identical to that fitted to the 8C 2300 sports car. The engine would run well on a petrol/benzol mixture, or alcohol could be added to choice, but Jano considered it vital that drivers did not allow the engine to exceed

5,500rpm, and for this reason two rev-counters were fitted in case one failed. Laurence Pomeroy wrote of the P3:*

> . . . never before had existing standards of performance been raised by so much in so short a time.

All in all, the P3 was Jano's most successful Grand Prix car, and an entirely original product of his genius.

In 1933 Alfa Romeo were in liquidation, and were nationalised. The new owners withdrew from racing and would not allow Ferrari to use the P3s until the end of the season, when his Monzas were being out-classed. Up to that time, the racing Alfas always carried the traditional *quadrifoglio*, or four-leaf clover emblem, first seen on Sivocci's 1923 Targa Florio-winning RL. From 1933 onwards, the Ferrari Alfas carried the famous Scuderia Ferrari Ravenna horse badge, and Alfa Romeo employees were forbidden to enter competitions.

Jano was a great admirer of Tazio Nuvolari, who he felt was a kindred spirit and a better driver than his great rival Achille Varzi. Before the 1934 Mille Miglia, Jano was in a particularly anti-Ferrari, anti-Varzi mood (Ferrari favoured Varzi rather than Nuvolari), and broke his company's rule by providing a specially-prepared 2·3 litre Monza for Nuvolari to drive against Varzi, who was entered in one of Ferrari's bored-out 2·6 litre Monzas. Despite his slightly smaller engine, Nuvolari led Varzi until the latter went ahead and won the race on wet roads due to having special rain tyres fitted. Nuvolari was second.

In 1934 Jano produced a road car whose engine was the ancestor of the Alfa Romeo cars of today, the twin-cam 6C 2300A. These cars were successfully raced in Italy, particularly at Pescara. In 2300B form, they had all-independently-sprung chassis as early as 1935, but few examples of this model reached the United Kingdom.

Thanks to Luigi Fusi, we have a fascinating insight into Jano's daily life at Alfa Romeo during his happiest years there. He used to walk to work every morning, smoking a cigarette in a long holder. This walk took him about ten minutes. He usually arrived at the technical office at about 8 am, and the rest of his team, who were due in at 8.30 am, at about 8.15. At lunch time, he would collect a bare chassis or a test car from the testing department and drive himself home for lunch. He went back to work at 2 pm, after a two-hour lunch break, and worked until 7 or 8 pm. He worked on Saturdays, too, and almost every Sunday he went out testing cars—to Modena, Bologna or to the mountainous regions north of Milan.

* *The Grand Prix Car* by Laurence H. Pomeroy, Motor Racing Publications Ltd, 1949.

On these Sunday test runs he often took Fusi with him, also Attilio Marinoni and/or Battista Guidotti, both well known as Alfa Romeo engineers and racing and test drivers. There were usually three of them on these runs, but four went if there were enough seats in the car. The others used to meet at Corso Sempione 71, and Jano would come down to greet them, often bringing Francesco, his small son, to see them off. When Alfa Romeo were pioneering high-quality, hydraulic shock-absorbers, at a time when most similar components had a notoriously short life, Jano and his team spent many Sundays testing various settings and modifying and refitting units as necessary. As a result of this, from 1934 to 1940, almost all independently-sprung Italian cars used the Alfa Romeo shock-absorbers perfected by Jano. Naturally, they featured on what is generally acknowledged as the world's fastest pre-war sports car, Jano's famous 2900B design, which had a detuned P3 engine in an all-independently-sprung chassis.

Jano had a theory that he could design a successful GP car that was down on engine power in comparison with its rivals, but made up for this by superior road-holding. Once the German Mercedes and Auto Unions had got into their stride from 1935 onwards, their vastly superior engine power became too great for even Jano to combat, however. In the autumn of 1934 he designed his Tipo C monoposto chassis, which was the basis of the 3·8 and 4 litre '8C 1935' and '12C 1936' straight-eight and V-12 cylinder cars, which occasionally beat the Auto Unions but never the Mercedes. These had independent rear suspension of the Porsche type, and Jano's own design for independent front suspension.

The Fascist pride of the State-controlled Milanese firm was picqued when these monoposto cars were not proving invincible against the racing cars of Nazi Germany, and Jano was ordered to get the cars winning races quickly, but he was not given enormous sums of money to spend on development work such as the Mercedes team, for instance, had at their disposal.

His final Grand Prix design for Alfa Romeo was the handsome 4½ litre V-12 '12C 1937', with similar independent suspension to the 1935–6 cars, but a lower tubular chassis. Delays in getting components due to aero-engine manufacture having priority, meant that the car was raced without having been properly tested, and when Guidotti's 12C 37 retired with back-axle trouble in the Italian GP at Leghorn in September 1937, Jano's friends said, 'This will be the end of Jano.'

And so it proved. The Spanish designer, Wilfredo Ricart, was angling for Jano's job, and was favoured by Alfa Romeo's manager, Ing Ugo Gobbato. Incredibly, the Alfa Romeo directorate decided that Jano was 'too old' to be capable of designing successful cars any more, so he was replaced by Ricart, who was six years his junior. Jano left Milan, and returned to Turin to work for Lancia.

Jano's designs continued to be used for some of the Alfa Romeo racing cars, whilst the all-conquering Type 158 was the work of his protégé, Gioacchino Colombo, who had first come to Alfa Romeo in January 1924, as a draughtsman in the special design department for the preparation of the P2.

Jano joined Lancia in 1938 as Direttore Reparto Esperienze (head of the experimental department), responsible for both cars and commercial vehicles. According to Fusi, Jano was not very happy at Lancia. He felt that the Torinese resented the fact that he had left Turin for Milan, and he was never so at ease working for Lancia as he had been at Alfa Romeo. His interests were always in his work and his family— and especially his son. Tragedy intervened during the war when Francesco died of an illness when in his early twenties. Francesco was a student in engineering at the Politecnico in Turin, and had spent much of his spare time doing social work. He had also joined the partisans during the war.

After his death, Jano became even more withdrawn, and he seemed to have no really close friends at Lancia. He and Rosina lived at a house named Villa Francesco on the corner of Via S Francesco d'Assisi in Turin.

The Jano influence at Lancia was first apparent in the 3rd Series Ardea model. The 1st Series was already in production when he arrived at Turin, but he was invited to pass judgement on it and suggest improvements. These were implemented gradually on the 2nd Series and almost completely in the 3rd Series. The Ardea was a little 903cc V-4 saloon, produced up until 1950, which was never available in Great Britain.

When the Italians' war ended in 1943, the R.E. was reduced in size and only Jano was retained, becoming head of both the R.E. and the technical office. When World War II ended, he began working secretly on the renewed designs of the Aurelia. The Aurelia (1950–9) and the 1,090cc Appia (1953–63) were the only two production Lancias with which he was involved from the outset.

The Aurelia progressed from a modest $1\frac{3}{4}$ litre touring saloon to a $2\frac{1}{2}$ litre GT car capable of 110–15mph. Front suspension was by the traditional Lancia vertical coil spring system, but the rear wheels were independently sprung on trailing wishbones and coil springs. The clutch and four-speed gearbox were in unit with the final drive, rear brakes were inboard, and the drive shafts passed through the hubs to outboard universal joints. It was a great car.

Ing Dott Francesco De Virgilio was responsible for the V-6 Aurelia engine, but emphasises that Jano was his chief at the time and thus in charge of the project. Like so many natural engineers of his era, Jano never bothered to obtain academic qualifications and letters after his name, and his title of *Ingegnere* was an honorary one. Yet De Virgilio, himself a qualified *Ingegnere*, says that Jano taught him a great deal. He has enormous regard for what he describes as Jano's 'fantastic ability'. He was a designer 'in particular' rather than 'in general', and investigated and discussed every detail of the vehicles, and always tested them himself.

Although an Aurelia won the 1953 Liège–Rome–Liège Rally (Johnny Claes) and the 1954 Monte Carlo Rally (Louis Chiron), it was not prominent in racing until a special sports/racing coupé was evolved, which had an enlarged supercharged engine, trailing links and a transverse leaf spring at the front with inboard brakes, and a leaf spring replacing the helical coil springs at the rear. One of these cars won the 1953 Targa Florio (Maglioli). By November of 1953, the D24 sports/racing car had appeared with an unsupercharged 3·3 litre engine and a De Dion rear axle. D24s took the first three places in the 1953 Carrera Panamericana race in Mexico and also gained victories in the 1954 Mille Miglia (Alberto Ascari) and the Targa Florio (Piero Taruffi).

The D50 Lancia Grand Prix car of 1954 was, in Fusi's opinion, Jano's best-ever design, and the best of its time. It was lighter, lower and shorter than its contemporaries, and the 90 degree V-8 engine (unusual in GP cars up to that time, though Maserati had raced a V-8 pre-war) formed part of the space-frame. The fuel tanks were mounted in pontoons on each side between the front and rear wheels, to reduce the wind resistance set up by them, and this feature contributed to the tendency for the cars to be too responsive and prone to sudden breakaway. Stirling Moss has said that, had modern racing tyres been available in 1954, things would have been very different, and Jano paid the penalty of being ahead of his time.

The first car was tested by Cav Callio on the perimeter track of Caselle airport on a freezing cold morning in February 1954. Design had begun in the second half of August 1953. It took practically the whole of 1954 to sort out the handling problems and, as always, the sixty-two-year-old Jano was taking part in the test driving and circulating Monza in the D50s himself—'Not as fast as Ascari, naturally,' De Virgilio remembers, 'but fast enough.' De Virgilio has repeatedly emphasised to us his opinion that it was most remarkable that Jano should have been capable of designing a revolutionary car like the GP Lancia at the age of over sixty—not to mention testing it as well.

The D50s all retired in their first race, the 1954 Spanish GP in October. Two crashed in the 1955 Argentine GP and a third retired with engine trouble, but by March of that year many of their problems were sorted out, and the cars won the Valentino and Naples GPs. At Monaco, Alberto Ascari had a brake lock and slid into the harbour at Monaco when in the lead. A week later Ascari was killed in a Ferrari at Monza, and in a short time Lancia withdrew from Grand Prix racing as the company was in liquidation and had passed out of the Lancia family's control.

Meanwhile, in the spring of 1955, Ing Fessia (designer of the pre-war Fiat 500, and later of the Lancia Flavia) had joined Lancia, and Jano signed a three-year contract as consultant and head of the racing department. It seemed that the new owners of the firm wanted a 'properly qualified' man as chief of the R. E. and technical office in place of Jano.

The D50 cars were handed over to Ferrari on 26 July 1955, in the

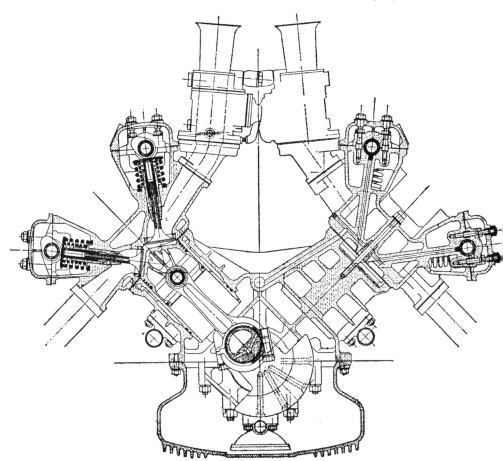

Cross-section of the D50 Lancia V-8 Grand Prix engine

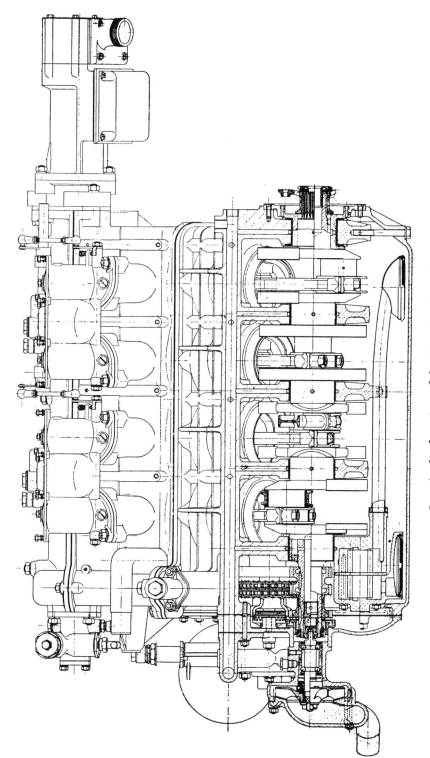

Longitudinal section of the D50 Lancia engine

courtyard of Via Caraglio where 'two groups of men faced each other, not exactly hostile, but cold and impatient for the sad ceremony to end'.

With racing finished at Lancia, Jano no longer had a proper job there, though his contract as a consultant had yet to run out. He then became consultant with Ferrari, and even when the development on what became known as the Lancia-Ferrari Grand Prix cars came to an end, Jano continued to travel to Modena two or three times a week to add his knowledge to Ferrari's various design programmes.

Memories are short, and Giovanni Canestrini has told us that after the war Jano felt very much alone and isolated and no one seemed to know or recognise that he was designer of the P2 and P3 and the great pre-war Alfa Romeo sports cars. He often went to race meetings in an unofficial capacity, and liked to position himself so that he could time all the drivers through the corners and compare their lines. He thought Luigi Musso was the most promising of the younger drivers, and likened his style to that of Varzi.

Fusi has told how Jano used to return to Milan at every available opportunity to seek him out, together with old friends like Guidotti. One day the three of them went to Como to meet Peverelli, the Alfa Romeo concessionaire there, and Jano bought a Giulietta. He was delighted with it, but told Fusi he was afraid to go back to Turin in a Milanese car.

Latterly, Rosina, his wife, became unwell and had difficulty in getting up and down stairs, so Jano designed a kind of chair-lift for her alongside the stairs in their villa in Turin. More serious was the illness of Jano's brother, who died of cancer after a protracted illness. For months Jano watched his brother dying, and vowed that he would never let Rosina go through the same agony as his brother's family if the same thing should ever happen to him.

As a result of this vow, on 13 March 1965, the quiet, likeable Vittorio Jano, a genius in his own field, died suddenly and quite needlessly in tragic circumstances. His obituary in *La Stampa* of 14 March ran as follows:

Comm Vittorio Jano, one of the most noted designers of racing cars, shot himself yesterday morning at his home in Via Fratelli Carle, no 12. He was seventy-four and had been ill for several weeks—nothing serious, a type of bronchitis, but he was convinced he was suffering from cancer.

He had never shown the slightest intention of ending his life. Yesterday at 7 am, while his wife, Rosina, and the maid were preparing breakfast, they heard an explosion coming from the direction of the bedroom.

Signora Jano was the first to enter the room: her husband was lying on the bed, bloodstains covering the pillow. Attempts at first-aid were useless: Comm Jano had exploded the pistol in his mouth and death had been instantaneous.

9 GABRIEL VOISIN

by H. R. KOUSBROEK

GABRIEL VOISIN (Born 1880). French: Pioneer aeroplane designer and constructor. Switched to car manufacture after World War I. Noted for imaginative and spectacular aerodynamic bodywork. Proponent of Knight double-sleeve valve engines, much modified, with four, six and twelve cylinders, and unconventional transmission systems.

GABRIEL VOISIN

I AM NOT much given to topographical nostalgia, and yet, having completed my notes for this essay, I felt that there was one thing left to do before starting on the actual writing. So on that particular December afternoon, I drove to Issy-les-Moulineaux and was prepared to feel quite prosaic as I stepped on to the bleak pavement of the Avenue Gambetta.

This rather ordinary-looking avenue is bordered on one side by a vast open space, covered mainly with sparse, moth-eaten grass, and traditionally known as the *champ de manoeuvres*. In fact, it has served as the heliport of Paris in recent years, and one has to go back to maps dating from the turn of the century to find it actually marked as *champ de manoeuvres*. At one moment during that era it was the site where, for the first time in history, a man-made machine was officially observed to leave the ground under its own power, fly a kilometre in a narrow loop, and land again at the spot it had started from. The *champ de manoeuvres* has remained an airfield ever since, in fact there can be little doubt that it is the oldest airfield in the world.

At that time—on 13 January 1908—none of the buildings which were later to line the opposite of the avenue existed. The two-storied wooden structure that became No 36 Avenue Gambetta went up in 1910 and the last extensive modifications were made in 1934. Today nothing remains. No 36 was razed to the ground in 1964 and the remaining buildings followed a few years later.

Even though they had survived, what would not, of course, have been seen—hence the pointlessness of such expeditions—was their history, any more than the *champ de manoeuvres* across the road bears the traces of what occurred there 30ft above the ground sixty-two years ago. And yet history is equally present on both sides of the avenue. He who goes there to chase shadows has to provide them himself: and shadows there are in plenty, for through the gates of No 24, which for years bore the remains of a decoration in the form of a stylised pair of wings with traces of blue paint, rolled a prodigious cortège of 10,400 aeroplanes and 27,200 motorcars.

Besides several thousand engines of the most varied description, prefabricated houses (produced in 1919, and, so far as I know, antedating any other such constructions), inflatable aeroplane hangars, and even an entire theatre seating more than two hundred people, which could be inflated in a few minutes by a $\frac{1}{4}$hp motor-pump, originated in these buildings. Quite a few beautiful women were made there as well, and although that must remain a different subject, they were very much a part of the ghostly ballet of people and machines that the imagination can conjure up in the Avenue Gambetta at Issy-les-Moulineaux.

*   *   *

I first met Gabriel Voisin in the summer of 1961. He was then eighty-one years old (he was born on 5 February 1880), and in poor health. 'If you do not come to see me soon,' he had written, 'you will find but an empty shell devoid of thought'. For a man who has lost a great deal in the past, the only loss Voisin ever complains about is the loss of his mental clarity. If this is true, as it must be, then his mental capacities in his best years must have been prodigious indeed. It is by no means easy to persuade Gabriel Voisin to talk about his past. He can, of course, be made to reminisce, but he will not readily fight old battles or get involved in problems long since solved. Even when I met him again later, in 1964, Voisin was still living in the future and it was only the technological challenges of our own day that really interest him. He is one of the few great minds in the field of technology whose genius did not lie in the elaboration of a single brilliant concept, but rather in his capacity to adopt completely new frames of reference, and then to imagine new possibilities in those terms.

It is difficult to realise that Gabriel Voisin was forty years old when he started to design automobiles, and that he had already terminated a career that would have given him a place in history even if he had never done anything else: his career in aeronautics. As if this were not enough, he embarked on a fresh venture at the age of seventy. Dispossessed of his factory and obliged to start again from scratch, he produced a means of transport of entirely different conception, bearing not the slightest resemblance to his automobile production of between the wars, in the form of the Biscooter, and I think that if he were given the means, he would even now come up with something novel.

Practically everybody in Tournus knows who Gabriel Voisin is, and we had no trouble in getting directions for La Cadolle, where Voisin was living at the time we first met him. La Cadolle is a low, modest building, situated in great isolation on the bank of the river Saône, a little over three miles from Tournus. The house is surrounded by a tall hedge, through which I could distinguish a big trailer, a pair of Biscooters, a 13-CV Voisin Type 'C' 15 of 1928 or 29, and a 2-CV Citroen. Strangers were invited to keep out by several large inscriptions in runny red paint, indicating the presence of *pièges à feu*. These devices, which play a prominent role in the rural sociology of France, are genuine booby-traps consisting of a small 16mm cannon, triggered by a string cunningly arranged to greet trespassers with a discharge of gunpowder, to which provident souls occasionally add a charge of coarse salt, since the use of lead shot in such devices happens to be illegal.

These inscriptions did not exactly produce an atmosphere of welcome, nor did it help much that Voisin had forgotten about our appointment, or at least failed to identify me. The misunderstanding was mutual: it never occurred to me that the person who emerged from the garden might be Voisin himself. In France, there is usually a connection between clothing and status, and we took him to be the gardener come to defend his roses. His small, stringy body swam about somewhere inside a sloppy blue overall. On his bare feet were ancient espadrilles. Over his arm was a great rush basket, and on his thin white hair, a faded blue beret.

But the manner of address when he did speak soon made it clear that we were in the presence of what had once been Gabriel Voisin, the dashing young man-about-town of the twenties at their wildest, owner of a string of ballerinas and of a legendary Paris mansion into whose swimming-pool the bright young things of the day pushed each other fully clad in spangled evening dresses, the elegant industrialist who rubbed shoulders with presidents and offered Mediterranean cruises to a dozen guests in his regal yachts.

It was still quite easy to understand why this man has been surrounded by women* all his life: my own wife immediately fell under his spell in spite of his great age. Clearly, there was more to it than the charm of his Edwardian manners. Late in life he still married a pretty dancer half his age. She and her equally pretty sister, there in the middle of nowhere on the banks of the Saône, have turned his house into a kind of boudoir. These immaculately turned-out creatures (for whom? for themselves) lead a life of Japanese court-ladies among their antiques and their embroidery, surrounding Voisin with little attentions, now and then briskly driving off to go shopping in their 2-CV.

Although one cannot say that Voisin has fallen upon evil days, he has lost his factory and most of his fortune, and it was thanks to his own generosity during the good years that he now lives in this cottage, which he had once bestowed upon a lady he much admired. Many years later, having prospered in life, she was able to offer his gift back to him.

---

* 'I have always enjoyed the company of women. I appreciated their intelligence, their subtlety, their resistance, their abandon. I loved to enjoy their joys, share their difficult existence, and I was well aware that within a few milennia all this masculine nonsense, this absolutely irremediable obtuseness, these pretensions, would disappear from creation, for the spirit has always been the victor in the struggle for life.'—(*Mes 1001 Voitures*, by Gabriel Voisin, p 41.)

It was here that he settled after having wound up the Biscooter contract in Spain, and where he wrote his memoirs. There are two volumes
of these: *Mes 10,000 cerfs-volants* and *Mes 1001 voitures*, published in
1960 and 1962 respectively. An excellent English translation of the first
volume, which deals with Voisin's aeronautical years, was published in
1963 by Putnam under the title *Men, Women and 10,000 Kites*; the second
volume, dealing with Voisin's automobile career, unfortunately remains untranslated. Next, came a book of hunting souvenirs, *Nos éton-
nantes chasses*, published in 1963 by La Table Ronde, with illustrations
by Voisin himself. There remains quite a bit of yet unpublished material
that might be characterised as socio-philosophy, the fruit of Voisin's experience, his suggestions for the future: for a more rational, less taboo-
and ignorance-ridden future—much of his present social thinking is as
revolutionary as were his pronouncements in the depths of the Depression, when he startled his entourage and infuriated fellow-industrialists
by preaching the virtues of shorter working-hours at undiminished
wages as a way of solving the current economic catastrophe.

Voisin has recently acquired the ruin of an old mill further inland.
He had become impatient with the crowds of Sunday fishermen swarm-
ing along the banks in front of his house, and with the farmer who, for
a tiny profit, had sacrificed a great semicircle of poplar trees that stood
at the edge of Voisin's property and masked it from the road.

This mill, which dates from the seventeenth century, has walls 2ft
thick and is being restored from scratch, with Voisin, who was originally
trained as an architect, not only directing the work but doing much of it
himself, despite a cardiac condition. Voisin has had two brushes with
death, and there are always oxygen cylinders in the house; at the time
of our first visit he was convalescing from one of these attacks and was
both physically and mentally in much better shape than he had been a
few years previously.

*   *   *

The most important of Gabriel Voisin's three 'lives' is definitely the
first, and he evidently has more of an emotional investment in his aeroplanes than in his cars. He speaks of his cars with a certain detachment,
when one can get him to speak of them at all, especially the ones of the
vintage and post-vintage period. He seems largely indifferent to opinion
about them, and wonders why anyone should be interested. 'I was blind
to a number of basic problems in the concept of the automobile in those
days,' he says, and he clearly considers his Biscooters more worthy of
interest. But his aeroplanes must have been the deepest involvement of
his life; they were also its tragedy.

Gabriel Voisin started experimenting with Hargrave-type gliders as early as 1898. Clear photographs exist of these, showing the continuity of design, through the subsequent models tried out at Berck beach and on the Seine, to the powered Voisin plane that succeeded in flying 260ft on 10 March 1907. After a year's further development of this machine, the kilometre in closed circuit was achieved on 13 January 1908—a performance involving independent take-off on the level, manoeuvres in pitch and azimuth, and landing, and since all this took place under strict official control, the crucial part of the claim that this performance was the first really independent, powered and sustained flight does not consist so much of proving this to be true as of proving that earlier flights were not independent, powered, or sustained. The only other contenders are, of course, the Wright brothers, and the machine with which they first flew at Kill Devil Hill on 17 December 1903.

As C. H. Gibbs-Smith has written,* all claims opposed to those of the Wright brothers have generally been advanced because of a failure to appreciate reasonable criteria by which powered flying should be judged, because of partial or complete ignorance of what the Wrights achieved, or of some discreditable intention to denigrate the Wrights.

As for the first argument, in Voisin's view the criteria by which the Wrights' flight are judged are indeed not applicable to powered flight, since he thinks that the Wright plane was a motorised glider. His main arguments are that it was dependent on some form of launching-device and that it was incapable of flight in the absence of certain air currents. The Wrights' Flyer No 1 of 1903 was launched from a sort of chariot on rails, apparently pushed by human agents, and all later models were launched by means of a weight and derrick catapult device. From descriptions of the behaviour of the Wright plane in flight, Voisin draws further evidence for the contention that it was basically a glider, and that the Wrights had not, in fact, solved the problem of powered flight.

There is also the question of the engine: the original American engine of the 1903 Wright appears to have weighed 240lb and to have delivered only 12bhp—0·05bhp per lb, while the 'Antoinette' that equipped the Voisins produced 50bhp at a weight of 145lb—0·3bhp per lb.

As for partial or complete ignorance of what the Wrights achieved, in Voisin's view this is precisely one of the main difficulties. Just what did they achieve? Voisin points out that not one of the Wrights' flights prior to 1908 took place under any form of inspection or supervision. There are, of course, eye-witness accounts, but these amount to little more than statements that the machine flew, which is not what Voisin contests; they are moreover contradictory.

* *The Aeroplane, an Historical Survey*, H.M.S.O., 1960.

Voisin himself adds a further criterion, ie, the inherent capacity for development of a given design, or the extent to which it determined the trend of later evolution; he notes that aviation has developed along the patterns set by the French designs of 1908 and owes little to the Wrights. No Wright, or anything like it, participated in World War I, which saw 150,000 aeroplanes in action, including 10,000 Voisins.

Though Voisin feels that recognition of certain things he has done has been maliciously withheld, it cannot be said that he is an embittered man—his extraordinary capacity for living in the future and throwing himself into new ventures has saved him from that. But he has been trying to get a hearing for years. He has published numerous articles, several illustrated and documented pamphlets, and, more recently, three hundred pages of memoirs about the early days of aviation, which he hoped would revive interest in the question.

This turned out to be the final disillusion. 'I received stacks of letters,' Voisin once told me sadly, 'but almost none were about aeronautics. The only thing that seems to have aroused any interest was my love-life.'

*   *   *

It would, of course, be unrealistic to suppose that all Voisin's difficulties sprang exclusively from the ill-will of others, and it must be said that quite a few of his controversial feats were already surrounded by an atmosphere of antagonism before they had taken place—a climate he had often helped to create. It almost seems, sometimes, as if Voisin had a tendency to treat people as imbeciles even before they had had a chance to prove that they were; and then, when they did, Voisin would note with wry triumph that he had been right as usual.

The various items of technical literature accompanying his products often make curious reading. 'We have divided this manual into two distinct parts,' begins the instruction book of the 13-CV, 'the first, short, incomplete, will perhaps be read; but the second, which represents an enormous amount of work, will doubtless never be opened.' After this introduction, Voisin takes the unsuspecting new owner to task for all the ways in which he can, and surely will, ruin the piece of machinery he has just acquired. 'Then, if anything goes wrong, the moment has come to blame the manufacturer,' he notes resentfully, and by the time he gets down to the actual maintenance instructions his tone has become cynical and reproachful: 'If the owner of VOISINS insist upon continuing such abusive treatment, which we have no means of preventing, we might perhaps request them to do something they rarely do: SERVICE THEIR CAR AND ABOVE ALL NOT OVERSTRAIN THE MACHINERY.'

The Voisin catalogues, especially those from before 1930, are veritable pieces of pamphleteering in the best tradition of the eighteenth-century French polemicists. What he fought was the hopeless incompetence with which people in the machine age use machines, and the absurd criteria by which they assess their merits. Advertisements that did not refer to concrete considerations incensed him. 'We have before us,' he wrote in 1927, 'some of the more beautiful examples of ridiculous flummery that certain foreign (read: American) firms have brought into fashion: "Like the eagle that dominates the skies . . .", "The mountain-goat scorns steep slopes and abrupt descents", "The greyhound, built for speed", "The courage and strength of a panther", and so many other idiocies from which one can deduce neither the horsepower of an engine nor the shadow of any information concerning petrol consumption. . . .' His own advertisements always appealed to the rational faculties and for many years his firm ran an advertisement offering 500,000Frs to anyone who would present a car of the same horsepower that could equal the timed performance and measured efficiency of the current Voisin models. There were no challengers.

Voisin broke the rules of accepted commercial etiquette by placidly listing all the defects his careful tests had revealed in other makes, while gratefully using the high-flown epithets of 'eagle', 'mountain-goat', or 'panther' to identify the brands. Year after year, and with diminishing patience, he tried to induce people to apply rational criteria when buying automobiles. 'Take a piece of paper,' he pleaded, 'and note on it what purposes the car must serve, how many people it must transport, at what speed, and with how much luggage. Obviously someone who wants to travel with six passengers and their luggage, at an average of 40mph at a cost of 40 centimes per person/mile needs a different car from someone who wants to transport two people at a 60mph average, and who is willing to pay 4 francs per person/mile. Both cars exist, but the buyer must not expect one to do the job of the other.' In order to provide buyers with the basic elements of choice, Voisin regularly published the detailed running costs and conditions of each model sold, computed from the officially controlled figures of actual distance-runs. The purchase of the faster models and sports cars included a trial at Montlhéry, timed in the presence of the customer, if he was available. Voisin gave prospective clients a set of four commandments: '1. Decide what purpose the car is to serve. 2. The opinion of a friend is worthless. 3. Publicity that does not give facts is a despicable imbecility and a waste of your time. 4. A car that weighs more than 40lb per bhp should be rejected, no matter how advantageous the price.' And, also from the 1927 catalogue, as an afterthought: 'The purchaser of a motorcar has always reminded us of a child wanting to buy a toy.'

That may not be a diplomatic statement, but it implies that Gabriel Voisin was one of the very few, if not the only automobile manufacturer, who refused to treat buyers as children, and who was not prepared to sell them toys. He has consistently fought the 'morbid fog of a criminal sentimentality' in people, instead of encouraging, flattering and cultivating it as everyone else did—and does.

*   *   *

Most of the great personalities in automobile history were basically designers of engines, with or without some outspoken ideas on chassis and suspension. But to Gabriel Voisin, as is evident from the foregoing, a motorcar was the whole thing, defined by the purpose it had to serve, and each component was designed with equal artistry in terms of its function. Thus Voisins have always been among the fastest cars of their time, but not because Voisin pursued speed for its own sake; speed was, so to speak, an inevitable by-product of the quality of the overall design.

There have been Voisin sports cars, and even very memorable ones, but the term that applies naturally to every Voisin, including the Biscooter, is—'high-performance car'. Characteristically, the kind of test or competition in which Voisins have always shown up to advantage was those in which the machines had to perform under normal road conditions—for this was what they had been designed for.

One of the first achievements to get much publicity was the Paris–Nice run made on 6 April 1921 by Dominique Lamberjack in an 18-CV. The Paris–Nice route was the classic itinerary for such demonstrations; the best time made before World War I was 16hr 15min, by Sorel in a 60hp De Dietrich. After the Armistice, when automobile production in France started to regain momentum, interest in the Paris–Nice run revived. The first to undertake it, as far as I know, was Louis Delage. It took him 14hr in a 4·5 litre, six-cylinder ohv machine known by the gaseous name of 'CO 2'. Soon the heat was on, and shortly afterwards André Dubonnet, in the new 32-CV (45hp) Hispano—the same that won in Boulogne later that year—reduced the time to 12hr 55min, representing an average of 45·5mph. It gives one an idea of the exceptional qualities of the 18-CV Voisin, that with its cylinder capacity of 4 litres (as compared to 6·6 litres of the Hispano), it was able to knock more than an hour off Dubonnet's time: 11hr 30min. This works out at 51mph, which even today is a quite impressive average, considering that the second half of the course runs through the Alps. Several attempts were made to better it, but not until 1923 did Baehr and Bloch, also on a 32-CV Hispano-Suiza, manage to better Voisin's time by six minutes.

Page 265 (*left*) Gabriel and Charles Voisin, aged eight and six; (*right*) Gabriel and Charles working on the engine of the plane in which Charles flew 260ft in 1907; (*below*) Henry Farman, Alberto Santos-Dumont and Gabriel Voisin in 1933 in front of the monument commemorating the kilometre flight of 1908

Page 266 (*above*) The first Voisin: Gabriel Voisin at the wheel of the 1899 car; (*below*) Rudolph Valentino and his wife in their 18CV Voisins in Paris, 1923

Page 267 (above) The start of the 1929 Strasbourg Grand Prix; the car on the right is a Bignan; (below) the Voisin 8CV Type C4 of the Paris–Milan run

Page 268 (*above*) A Voisin 4-litre 'Laboratoire' of 1924; (*below*) Voisin GP car competing in the 1923 French Grand Prix at Tours

The last time a Voisin could be seen on this historic itinerary must have been in 1960, when Voisin himself, at the age of eighty, drove a thirty-year-old 17-CV (Type C 23) from Paris to Nice. 'On the long *route du Soleil* there were few cars able to overtake us with ease,' he notes in his memoirs, 'and despite the intolerable traffic conditions I covered the 600 miles in 14 hours.'

The praises of the 18-CV Voisin have been sung before; it quickly became the darling of the early twenties, a favourite at *Concours d'Élégance*, and everybody who was anybody had to have one. Anatole France and H. G. Wells, Josephine Baker, Maurice Chevalier, Raimu, and many now forgotten celebrities of the time drove Voisins; Rudolph Valentino had two, and Millerand, President of the French Republic, four. Charles Faroux called the 18-CV Voisin the best four-cylinder car in the world, and in his view it was because of its success that a number of other makes adopted sleeve valves at the time. A case in point is of course the 18-CV Peugeot, which was a faithful copy of the Voisin, all the more so for having been based on the same drawings by Artauld and Dufresne, from which the 18-CV Voisin had been developed.

What Voisin had done to the sleeve-valve engine may be gathered from the fact that only Hispano-Suiza slightly surpassed its efficiency of 20bhp/litre, while it should be noted that peak horsepower on both engines was deliberately sacrificed to obtain power in the middle range. According to a study that appeared in *La Vie Automobile* in April 1921, no other engines could come near this figure, and although that may not be completely true, up to 1950 machines of similar capacity existed (such as the Pontiac 25), that produced less bhp per litre than the 18-CV Voisin of 1920.

Nor did matters remain there: from 80bhp in 1920, the power of the 18-CV rose to 120bhp in 1922 and 140 in 1924, while even higher figures had been obtained on the bench. The 17-CV Voisin Type C23 of 1931 was said to have been the first unsupercharged production engine to yield more than 100bhp from three litres—and in 1963 Gabriel Voisin could still be seen working on a design that had to produce 100bhp per litre; but although that was also a 'valveless' engine, it was no longer a sleeve-valve (it was a two-stroke). From 1919 to 1937, when he abandoned automobile production, Voisin was faithful to the sleeve-valve; he even constructed radial engines with sleeve-valves. When I asked him if this was through preference or because the tradition had become an obligation, he replied that he has never liked poppet-valves, and he believes even now that the two-stroke 'valveless' has more of a future than the poppet-valve. So as far as I know, Voisin designed only two poppet-valve engines: a 500cc single-cylinder for the 1919

'Sulky' cyclecar, and a fantastic forty-two-cylinder (in six banks of seven) 2,000hp aero-engine designed in 1939. They had single overhead camshafts and inclined valves. There may have been other aero-engines.

In the vintage era, sleeve-valve engines, in Voisin's view, had certain real advantages over other systems. Apart from silence, which was essential to Voisin's conception of the automobile (viz the memorable statement in a certain Voisin advertisement of 1927: 'Of all sensations experienced by the human organism, noise is by far the most upsetting'), they presented the advantage of a desmodromic distribution, of inlet and exhaust areas equal or superior to the cylinder bore and of an ideally-shaped combustion chamber without hot spots, which explains why un-heard-of compression ratios such as 8:1 could be used in 1922. As a result Voisins were very easy on fuel.

They already had an advantage in this domain because of Voisin's tyrannical rules about body weight, which also accounted for longer tyre life. Voisin went so far as to refuse to guarantee his chassis if they had been fitted with the heavy coachwork that was the pride of con-temporary coachbuilders; their undying hatred from that time on-wards, and that of the financial powers behind them, almost caused the downfall of the firm at one point.

By January 1923, when the *Salon de l'Automobile* was held, Voisins had raked in ninety-four first prizes, fifty-nine cups, and a number of absolute best times for hill-climbs, including the Mont Ventoux, the Gaillon, the Turbie, the Alpilles—and, of course, they had taken first, second, third and fifth places in the 1922 Grand Prix at Strasbourg—the germ from which another great controversy was to spring. The silver cup that was won on that occasion is kept in Voisin's dining-room.

The Strasbourg GP was won because of the light weight and the favourable aerodynamic characteristics of the narrow *torpédo*-type bodies of the Voisins. Although it is almost impossible to imagine this today, at the time there were people who considered that to be an unfair advantage. They were even sufficiently numerous and influential to make the ACF change the rules for the next Grand Prix: for production cars, streamlining would thenceforth be forbidden and only reactionary box-shaped bodies—outline drawings were provided with the entry-regulations—could participate.

When this became known, less than six months before the date of the race, Voisin flew into one of his most memorable rages. He fired off an irate letter to the President of the ACF, in which he indicated that he would enter his cars in the racing car category instead.

What he meant was nothing more nor less than that he was going to enter production cars as racing cars; the point he was trying to make was that the performances of a motorcar could be drastically improved by streamlining and that a production car, if it were equipped with a suitably light and aerodynamic body, might even cut a reasonable figure against a racing car.

His fundamental error was, of course, to think that such subtleties would be generally understood, and the way he went about it did not help either, Voisin being far too aggressive to realise that men must be taught as if you taught them not. The only aspect to which people proved receptive was that Voisin had once again created an incident, that he had accused a number of highly-placed persons of holding idiotic opinions, and that he had, with much ado, entered four 'racing cars' in the Grand Prix de Vitesse. When he finished fifth, it was felt that he had failed. No wonder, with such new-fangled, bizarre contraptions.

This last aspect alone was remembered. That they represented reason and progress, as compared with the prescribed biscuit-boxes of the production GP car, was not appreciated. Even serious automobile historians have seen nothing more in the 1923 Voisin aerodynamic bodies than an opportunity to introduce a light note into their prose: '(the body-work) was supposed to resemble a section of an aeroplane wing, although the resemblance was perhaps not immediately apparent to the uninitiated'—in the words of someone who cannot generally be accused of uncritically copying unreliable information.

Another well-known book refers to 'one-dimensional [sic!] aerofoil bodies'. I have not been able to find anything to support this assertion in any original document, and Voisin himself says that it is nonsense. It must either originate from some translator's failure to distinguish between the terms 'aerofoil' and 'fuselage', or from some misinformed commentary of the time.

The fact that the Voisins of the 1923 GP de Tours were basically production cars is beyond any doubt. Mechanically, the car was entirely constituted from elements of the four-cylinder, 8-CV Type C 4, which had only just entered production. Sleeves, pistons, distribution, connecting-rods, clutch, gearbox, back and front axle, steering and brakes were absolutely stock. The only modification was the addition of two cylinders to reach the capacity of two litres. It delivered 65bhp (according to J. Rousseau, no more than 60), which was just about what could be expected from a good production engine in special tune.

The winning Sunbeams produced 102bhp. The winning speed was 75·3mph, while Lefebvre's Voisin, which finished fifth, had achieved 63·2mph. The winning Sunbeam, in other words, was 19 per cent faster with 57 per cent more horsepower; of course the relation between speed and horsepower is not linear, but these figures do illustrate the exceptional qualities of Voisin's aerodynamic coachwork, while they do not even take into account that a proper comparison should not necessarily be in terms of the winning speed, but of the *median* speed of the cars that finished. It is also interesting to know that Rougier's car, until it withdrew after 250 miles, had been doing 81mph.

All this verbiage to prove what is perfectly obvious: that the aerodynamic, lightweight (84lb), enveloping body that Voisin had designed and made in a little over a week—the first monocoque body in the world—must be seen as a major achievement instead of the faintly interesting eccentricity as which it is usually described. Especially in view of the fact that aerodynamic, lightweight, enveloping monocoque bodies have not exactly proved to be without sequel in automobile history.

*   *   *

From the Tours GP cars Voisin developed a new sports model. Until then the official sports car of the marque had been the C 5, which was basically the car that had won the previous year in Strasbourg. The new sports cars were known as 'Laboratoires'; the lines of the Tours GP car are clearly visible in their shapes, and they must have been the sportiest machines that Voisin ever offered to private buyers.

The 18-CV engine, which developed 100bhp in C 5 form, was not anywhere near the limit of its possibilities. In the Strasbourg cars, it was said to deliver 120bhp, and shortly before that it had even reached the figure of 150bhp (at 4,000rpm) on the bench, through what was originally an error in the size of a set of experimental magnesium pistons. Because of this error, the compression ratio had risen from 5·5:1 to 8:1, and, contrary to current belief, it turned out that this did not cause spontaneous ignition. Voisin attributes this to the absence of hot spots in the sleeve-valve design. This high-compression engine, fitted with a reinforced crankcase and crankshaft (causing some of the newly-found horses to be lost again, incidentally), became the power-plant of the 'Laboratoires'. The coachwork, designed along the same lines as that of the Tours cars, was also of monocoque construction, but did not envelop the rear wheels and looked generally less forbidding. The 'Laboratoires' appeared towards the end of 1923 and their officially-timed top speed was 110mph.

The story of the 8-CV Type C 4 runs an almost parallel course. Like its big brother, this 1,240cc, four-cylinder machine first distinguished itself on the Paris–Nice itinerary. By 1923 this run had become an organised *critérium*, as it was called, in which even foreign makes participated, and the entire trip had to be accomplished with a sealed bonnet.

On 17 February 1923, the small car won the event outright, ahead of a Peugeot, a Lorraine-Dietrich, a Bentley, and a Chenard et Walcker.

A sports version was produced—a very pretty aluminium torpedo with striking horizontal mudguards—which won the Circuit des Routes Pavées and undertook a (quite illegal) race with the Orient Express to Milan, beating the train by more than two hours. It was on this car that Voisin first installed an overdrive—the first independent overdrive (of the kind that is distinct from the gearbox and provides two final drive ratios) in the world* : it consisted of an epicyclic gear train on the propeller shaft, operated by a mechanical lever. On later cars its operation became pneumatic (known as the Voisin 'compound relay'), to become electric in its final form. The bore of the 8-CV was soon enlarged, bringing the capacity to 1,328cc, and finally enlarged further to become the 10-CV Type C 7 of 1,551cc.

This 10-CV, which produced 44bhp at 4,000rpm in standard form, was used to power a diminutive replica of the 'Laboratoire', endowing it with a top speed of 81mph, as usual, officially timed. There has been considerable speculation about the small external propeller that characterised the Tours GP car and its derivatives: it drove a water-pump.

Four 'Laboratoires' were entered for the 1924 production-car GP at Lyons. The regulations no longer excluded streamlining; the 1923 production-car race with prescribed coachwork had been likened to a testing of engines that might have been better achieved on the bench, and the best times were under the winning times of the Voisins the year before.

Three of the competing 'Laboratoires' were 'big' (4 litre) ones and the fourth was of the small variety, but with a 2 litre engine. It was this Lyons GP that caused Gabriel Voisin to withdraw from racing and switch to speed records 'where inanimate chronometers are one's only opponents'. For Voisin had once again become extremely fed up with the live ones; his cars had been penalised for trifles and there had been ill luck of all kinds. Although manifestly the fastest cars in the race, they failed to win and Voisin had to be satisfied with being second. He wasn't. To make things worse, Voisin understood very well that it would not do to complain yet again to the ACF about the rules.

I like to think that the 'Laboratoires' would have proved their superiority if Voisin had not withdrawn from racing at that time. Or would the sleeve-valve really have turned out to be the obstacle that one is inclined to think it was from today's point of view? As late as 1937 some people were still preparing a team of GP cars with sleeve-valves (Minerva—they never raced). According to the few people who have known the 'Laboratoire', there was nothing like it at the time, and

* The 1907 Rolls-Royce Silver Ghost had an overdrive *gear*, which is something else.

Voisin himself considers it to be the one of the three models of all those he built whose performance was really hair-raising—the second being the 12 litre V-12 speed-record car and the third the straight-twelve.

What has become of all these incredible machines? No one seems to know, and no information has so far led to a surviving specimen. The 8-CV of the Routes Pavées was allegedly stolen and sold to Renault, where it was taken to pieces to discover its secret. As for the Tours GP cars, André Pisart, a Belgian who used to race for Chenard et Walcker, describes in his book on the Chenard story how one turned up at the GP de la Baule, which was run on the beach. This may also have been a 'Laboratoire', for there are few people who do not confuse the two. At any rate, during the trials it was driven into the sea by a madman. Some others seem to have ended in ditches at minor races. Very few had been made, perhaps two dozen in all. The remaining ones must have disappeared during the war for, with their light alloy bodies, to a scrap-merchant they must have looked as if they were made entirely of bank-notes. May I be mistaken!

*   *   *

It is perhaps just as well, on the other hand, that Voisin stopped racing when he did, for racing was clearly not in his blood. Although he has personally flown or driven all the machines he has made, including the speed-record cars, at full speed, he was never tempted to do any com-petition driving himself. Nor did he pilot his aircraft in the countless pageants and demonstrations in which they participated between 1908 and 1913. In that respect his brother Charles seems to have been different, and history might have taken another turn if a motor accident had not ended Charles Voisin's life in 1912.

Gabriel Voisin's dislike of speed for hell's sake must be rooted in his conception of heaven. To someone who hates the sound of a Bendix starter pinion crashing into mesh (Voisin calls it 'the noise that disgraces the automobile' and all his own designs were equipped with dynamotors mounted on the end of the crankshaft), pushing a machine to its limits must have become painful just at the point where other people are beginning to get their kicks. He wrote in 1935 that speed was meaningful to him only in terms of the silent, flexible, effortless performance of machines that are furnishing not more than a moderate part of the work they are capable of delivering.

Under the pressure of French fiscal legislation, Voisin designed several cars with relatively small engines which had to furnish a maximum amount of work from a minimum capacity. And he certainly did a splendid job, as witnessed by the fact that he designed sleeve-valve engines that could function safely and continuously at speeds of over 4,000rpm, and by the circumstance that surviving Voisins, which are predominantly of this type (more of them were made), are still quite reliable today. Nothing would enrage Voisin more, incidentally, than people calling these cars 'under-powered' because they failed to understand the difference.

On various occasions Voisin has argued that taxation by cylinder capacity is opposed to progress, even if it has in a way contributed to the development of the internal-combustion engine, just as racing formulae, in his view, should not be based upon capacity but on petrol consumption. All this relates to a principle which is forever tacitly implied in Voisin's criteria: mechanical perfection can only be defined in terms of purpose or function.*

The ideal source of power in a motor vehicle, to Voisin, was steam, and he has called steam engines 'the best of machines'. He had owned several steam cars in his aviation days, and in 1920 he undertook to construct a steam car on the chassis of an 18-CV in order to gain more insight into what constituted the unique sensation of perfection that these early machines had given him, despite their crude construction.

Five different models of twelve-cylinder cars were the result of Voisin's successive attempts to approach this perfection in an explosion-propelled vehicle—('no noise, no smoke, no vibration, no gymnastics of hands and feet, a magical flexibility. Lightning acceleration in silence, and always that incomparable sweetness, that total impossibility to distinguish the number of cylinders . . .') but the ultimate satisfaction must have eluded him, for at one time and another he still designed the odd steam car. Odd is the word, and the strangest of them all was an experimental set-up in which a conventional internal-combustion engine was used as a gas-producer, the steam engine working off the exhaust. This hybrid did function but, of course, had no practical applications. Another steam car—with a double-acting twin-cylinder compound engine of 200 + 500cc, weighing 38lb and producing 15–20bhp at 100psi—was developed during World War II. This might actually have been an opportunity to re-acquaint people with steam vehicles, for production of cars with internal-combustion engines had come to a halt and what few remained were useless for lack of petrol. But by the time

---

* A corollary of this view is the notion of 'negative progress', as one can witness it today, Voisin likes to point out, in aviation; progress, as it is understood today means the development of *faster* aeroplanes at constant or slightly rising *cost*, while in Voisin's view progress would be the development of *cheaper* aeroplanes at constant or slightly rising *speed*. The point being, in other words, that it is more desirable to go to New York at half the price than in half the time.

Voisin had managed to organise a very small production line, the war had ended. The engine of the prototype is still in existence.

Work on a first V-12, Type C 2 of 7·42 litres, started late in 1919; prototypes existed in the spring of 1920, and by 1921 the car was completely ready for production. It was shown at the Salon and it is tantalisingly listed in the various catalogues and automobile annuals of the time, but it was never commercialised. Which is a pity, since it was the nearest Voisin ever came to his conception of the automobile during the twenties—that is to say, his conception of the luxury car for which a market existed in the twenties (an early 'Biscooter' having failed on the other end of the scale, as we shall later see).

It is even more regrettable that none of the three final prototypes has survived, for it must have been easily the most advanced design of the time. The compressed air braking of the first prototype had been replaced by a hydraulic system. Each wheel had a closed hydraulic circuit of its own, comprising a small pump which was driven by the wheel itself. Not only did this mean that braking required no other effort than regulating the output of the pump, but also that wheel-lock was impossible. It is this system that has given rise to the legend that the front-wheel hydraulics were fed through the axle: there was no plumbing visible. The equally revolutionary clutch consisted of two opposed turbines in oil, with an electromagnetic lock at the end of their travel—this was no doubt with an eye to the possibility of orders from heads of state, for it had been necessary to fit the presidential 18-CVs with air-cooled multi-plate clutches to withstand prolonged driving at a walking pace. The rear springs were hydraulically adjustable and the engine was a *tour de force* of light-alloy casting: all the manifolds were inside the blocks, and nothing protruded but the single carburettor, placed at one end in the centre of the V. Why was this car never put into production? 'It was too complex and therefore prohibitively expensive to produce in as small a factory as ours', says Voisin. He has told me more than once that his works were not really equipped to turn out any V-12s at all.

The second V-12, Type C 17—still the first V-12 to be made in France—appeared in 1929 and was followed very soon, apparently in the same year, by a third, Type C 18, of slightly larger capacity (4·86 litres instead of 3·85). One of the speed-record cars (there were five in all) was of this type. The V-12s are well-remembered in France, with their electric gearboxes and underslung chassis; but few, if any, seem to have reached England, if the fact that they are paid little attention in the current literature on vintage cars is anything to judge by. Yet their memory lives on in the Phantom III Rolls-Royce: Voisin put his

Page 277 (above) The 5-litre V-12 Voisin Type C18; (below) the last Voisin produced at Issy: the Aérosport C28 of 1935

Page 278 (above) An early Biscooter: the 500-cc cyclecar of 1919; (below) the transversally-engined fwd Biscooter of 1950

complete documentation at the disposal of a team of Rolls-Royce engineers who came to Issy in 1934. It would be interesting to know if their report can still be found in that company's archives.

The fourth V-12 was specifically designed for the speed records and figures in this list mainly to swell the numbers and to enable me to observe that the Voisin works had no device big enough to measure its horsepower. The fifth twelve was not a V, but an in-line, the legendary straight-twelve for which Voisin is still remembered today by many people who don't know much else about him, in keeping with the modern taste for believe-it-or-not items. But authentic facts about Voisin's dodecuplets have always been very scarce.

The straight-twelve was developed as a rejoinder to the Packards, Cadillacs, and other heavy cars that had managed to nibble at a corner of the European luxury market by 1935. Voisin first designed a straight-eight of eight litres capacity, and then, dissatisfied, tried one of the earlier V-12s in a light chassis. He meant these cars to be really usable at speeds of 90mph and over, and found that controlling them at such speeds was a task far beyond the capabilities of the average driver. He attributed this to the presence of heavy masses at too great a distance from the centre of gravity.

In November 1935, Voisin's drafting department started work on a straight-twelve whose engine came into the driving space as far back as the instrument panel. The aim was to get the mass of the engine as close to the car's point of gravity as possible, and at the same time to shorten the transmission. The engine consisted of two C 23s in tandem, the total capacity thus amounting to six litres. (See drawing, p 280.) A 12:47 back axle was driven via a multidisc clutch and a four-speed electromagnetic gearbox, moving the car at the rate of 27mph (43·3km) per 1,000rpm. The red line was at 4,400rpm, and the brakes were hydraulic.

The first prototype was ready in December, and with the enthusiasm that characterised Voisin's teams, testing and adjusting had been completed by April. The exceptional handling characteristics at very high speeds that had been projected had been achieved, and all contemporary comments on the Voisin straight-twelve particularly praise its road-holding, power and silence. The car was made in two lengths: wheelbase 10ft 6in, taking the four-seat *Ailée* coachwork, which was practically identical to the Aerosport (C 28), the coachwork that set a trend which still lasts today. The longer chassis had a wheelbase of 11ft 8in and carried the *Croisière* coachwork, vaguely reminiscent of a Rolls-Royce Silver Wraith, and designed for a more conservative public.

The straight-twelve, known as 'the fastest production car in the world',

Q

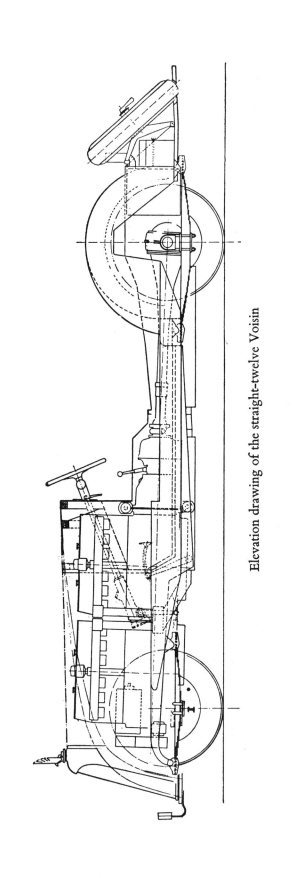

Elevation drawing of the straight-twelve Voisin

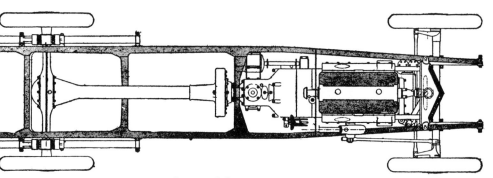

Chassis of the V-12 Voisin

was the last Voisin car to carry the well-known mascot, the swan-song of automobile production at Issy-les-Moulineaux.

Another car was indeed produced under the name of Voisin in 1938, but it is rarely understood that this machine was no brainchild of Voisin's. It was powered by a supercharged Graham Paige engine—a sorry piece of machinery, Voisin calls it—and it was concocted by a certain M. Garabeddian, who had bought up the remaining 'pontoon' platform-chassis, along with the brand-name Voisin, from a financial group which had managed to gain control of the enterprise.

As in 1931, Voisin had been 'neutralised', and once again the *phynanciers* very quickly succeeded in destroying the delicate and almost miraculous set-up by which the Voisin factory had been able to play a role that was in complete disproportion to its technical and financial means. It is hard to tell what the essence of that delicate set-up can have been but certainly it was no myth, it existed. And not only women fell under Voisin's spell—I came upon the traces of that spell when I spoke to people in Issy who used to work for him, and again in a letter to Voisin that André Lefèbvre wrote not long before his death. When,

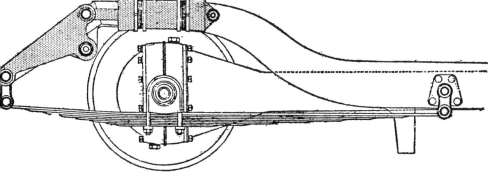

Unusual rear suspension of the V-12 Voisin

after the success of the speed records, several firms made offers to Voisin's principal engineers—Marius Bernard and André Lefèbvre—they replied that they had a sentimental attachment to the firm. It was at Voisin's own insistence that Lefèbvre finally went to Citroën.

I once asked Voisin what he would do if he got his factory back. 'I wouldn't want it back,' he said, 'and in any case, it would be necessary to resuscitate all those who were with me then.'

*   *   *

Unlike the way things went in 1931, Voisin never regained full control of his factory after 1937, and the slow process of decay, of which I saw the results in Issy-les-Moulineaux, may be said to have started at that point.

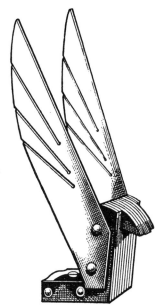

The Voisin winged radiator-mascot

The winged mascot, whose reign thereby came to an end, was never popular with Gabriel Voisin. Early photographs show that Voisin-owners mounted Rolls-Royce-type damsels on their radiator-caps, and it was early in the twenties that Voisin, at the request of a client, de-signed the riveted 'birdie' as it was subsequently called (the term 'birdie' carries almost all the meanings of the French 'cocotte', which is really a nursery word for 'hen'). 'I immediately drew the mascot,' Voisin wrote in 1930. 'I had the thing made and was soon dissatisfied with this bad joke. But despite all my efforts, the birdie has remained

firmly stuck to our filler caps ever since. The bad taste I had unleashed
had taken root. This story is applicable to all multi-chromed motorcars
that are adorned with totally useless accessories.'

Needless to say, the story that Gabriel Voisin himself solemnly
screwed the mascot in place after a chassis had been tested, as a sort of
official consecration, is apocryphal.

The bat-like design of the Voisin trademark and radiator emblem,
upon which the mascot was based, was not a bat originally. The emblem
dates from the aviation period, and was originally an Egyptian figure;
some checking confirms that it is an almost unmodified copy of the
winged Isis from the tomb of Seti I. It is an interesting thought that the
Voisin bat, as well as the Voisin radiator, in turn may have served as
models for the Bentley label and radiator, as has been observed more
than once.

This bit of micro-historiography reminds me of another curious
detail: the origin of the letter 'C', by which all Voisins produced in Issy
were designated. When Gabriel and Charles Voisin built their first car,
in 1899, they referred to it as *le char* (the waggon). (By a rare stroke of
luck, a photograph of this vehicle, of which there is a description in
Men, Women and 10,000 Kites has just come to light.) The 'C' of all
Voisin cars stands for *char*, as a tribute from Gabriel Voisin to his brother's
memory.

*   *   *

So dissimilar is the Biscooter to any of the Voisin cars discussed so far,
that one would almost think that it had been designed by a different
person. And yet the Biscooter has direct ancestors among some little-
known Voisin designs. The oldest of these dates back to 1919. At the
first Salon de l'Automobile after the war, the Voisin firm not only
exhibited its 18-CV, but also two tiny vehicles, described in the contemporary press as *engins démocratiques*. One was a bicycle powered by
a 125cc two-stroke engine, known by the English name of 'Motor Fly',
and the other a four-wheeled cyclecar that went by the equally anglo-
phonic title of 'Sulky'.

The Motor Fly did actually have something of a career; it even won
several races: the Circuit Normand, the Houlgate hill-climb, and the
Circuit de Choisy-le-Roi. As for the Sulky, the only positive item seems
to be that Lefèbvre drove one to Deauville and back. Speeds of a little
over 35mph were possible with the single-cylinder, 500cc, four-stroke
engine, placed off-centre near the left-hand-side rear wheel (invisible on
the photograph), driving the rear wheels through a three-speed gearbox. Earlier prototypes had been equipped with a variable-ratio friction-

drive transmission, which produced 'more sparks than torque' and was abandoned.

The other ancestor of the Biscooter was the Voisin 3 litre front-wheel-drive V-8 of 1931, which was never commercialised—as mentioned earlier, in 1931 a *phynancial* group (the *General Motors Européenne*) had gained control of the Voisin firm, and they would not have it.

Front-wheel-drive was on a number of people's minds toward the end of the vintage era, stimulated by, among others, Grégoire's Tracta. The fwd V-8 that Gabriel Voisin and André Lefèbvre designed and built is probably not the only fwd prototype that was built in France at the time, but it is certainly the only one that has the illustrious distinction of being the immediate ancestor of the well-known fwd Citroën, as well as the fwd Biscooter. It is still not very well known that Lefèbvre, who had joined Citroën on Voisin's recommendation when he could stand the *General Motors Européenne* no longer, is not only the father of the car known in England as the Light Fifteen, but also of the immortal 2-CV (designed in 1939) and of the ID and DS 19.

One of the features the Biscooter inherited directly from the front-wheel-drive Voisin V-8 was the single brake on the drive-shaft. Voisin had experimented with a centrifugally-locking differential on this V-8 (on the principle that a differential is an asset only at low speeds, and becomes a liability at high speeds) and in the end he had decided to do away with the differential altogether. The two brake-drums then fused into one.

The Biscooter's genesis is a subject in itself, and as the success of the Issigonis Mini has aroused interest in the ancestry of transverse-engined, fwd minicars, it is worth knowing that the Biscooter-Voisin was one of the first vehicles of this species to have been seriously produced. The principle was pioneered, I believe, by Walter Christie in 1904, whose models (which were moreover no minicars) remained experimental. The first real forerunner must have been the 'Micron' of 1925; not many of this astonishing single-seater rolled out of the factory in Castanet (near Toulouse), but still enough for some to have survived and to make it the first commercially-produced, transverse-engined fwd minicar. Long before DKW in other words—and also the space-saving layout of putting the gearbox in the sump underneath the engine makes it a very pure specimen of the animal under consideration. 'Microns' were designed by Henri Jany, who entered an even more striking version of the same idea in the 1934 Concours of the *Société des Ingénieurs d'Automobile*; but this design never went into production, causing Voisin's Biscooter, of which 40,000 were produced under licence by Autonacional in Barcelona between 1950 and 1956, to be the next ancestor in line, (if one agrees that DKW does not really belong to the same family (but rather to the Latil-Tracta-Citroën group).

Towards the end of the Vintage era, Voisin started experimenting with different locations for engine and driving-wheels. From that period dates what might be called a 'middle-wheel-drive' lozenge-shaped car (an idea that was revived a few years ago in the form of the Pininfarina XPF 1000), which Voisin claims he really intended to put into production. (See drawing below.) The final obstacle was not the sufficiently formidable problem of how to get the car over a service-pit, but of how to provide its radial engine with an acceptable idling speed. This remarkable, ultra-light, seven-cylinder radial engine of four litres had sleeve valves and fuel injection, and one of Voisin's hundred-odd patents (he did not believe in patents and allowed most of his inventions to be copied freely; his frequent comment when inspecting a car: 'they don't even know how to copy') concerns an ingenious silent injection-pump, which he designed for this engine.

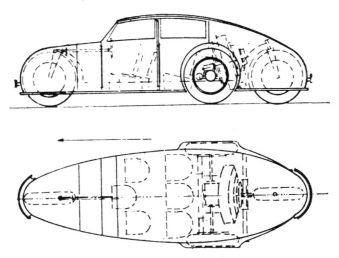

The lozenge-shaped, middle-wheel-drive Voisin

Besides experimenting with 'middle-wheel-drive', Voisin used the same seven-cylinder radial engine and transmission unit to test the possibilities of 'everything at the rear', and there is a photo of a conventional chassis with this unit mounted in the place of the back axle. But, finally, front-wheel-drive seems to have carried the day with Voisin, and he notes that in 1938 he still had hopes of putting the fwd V-8 into production. The prototype, incidentally, was bought by a Belgian and may now be languishing at some spa.

As for the transverse mounting of the engine, Voisin has told me that it did occur to him at the time of the V-8. And, of course, Voisin's first car of 1899 had both a transverse engine and gearbox, but this arrangement was not rare in Veteran days; in fact, Georges Richard's first car of 1899, the 'Poney', is strikingly similar in layout to Voisin's 'pigpusher' (see *Men, Women and 10,000 Kites*, p 90). By 1945, when Voisin designed the Biscooter, there was not the slightest question in his mind but that the engine was to be transversal and situated above the gears.

By the time the Biscooter was in production, Voisin had come up with more new ideas, such as steering on all four wheels by means of two coaxial steering-wheels that could be locked together or used independently; various applications of his theories on suspension and steering geometry; more innovations in coachwork and in the techniques of producing it cheaply, including the long-overdue sliding doors, ingenious arrangements in climatisation, which provided fresh air without draughts and allowed a car to be parked in the sun without becoming a furnace, and, on fast models, the use of stabilising surfaces and air brakes.

Voisin constructed a number of prototypes incorporating these innovations between 1955 and 1960, when he had a small drawing-office in the rue des Patures, in Paris. Among the things produced there were a steam tank, a light all-purpose amphibian car for the French Army, and a 50cc single-seater calculated to beat modern traffic and parking conditions as well as administrative complications and fiscal legislation (none of which apply to 50cc vehicles), with a target selling-price of under £50. A number of countries nibbled at an improved Biscooter licence, including India, Morocco, Israel, Ireland and Argentina (where a plan for the large-scale production of such machines was on the verge of adoption when the Peronista regime fell), but nothing ever came of it. Plans for the production in underdeveloped countries of a set of utility cars of 125 and 1,275cc had been worked out in detail, involving a bare minimum of machine-tools and skilled labour, but they were invariably eventually rejected by governments which felt that the cars lacked 'prestige.'

What reveals itself as the revolutionary aspect of the Biscooter and its derivatives goes beyond the transversal engine fwd principle; it is rather the use of this layout to change the whole concept of the automobile (cf *Mes 1001 Voitures*, p 213), definitively laying the spectre of the traction-animal in automobile design, and at the same time adapting private transport to the radically changed traffic conditions of today.

It is interesting to recall at this point, that, in the 'twenties, Voisin had already worked on a visionary plan for the city of Paris, in collaboration with Le Corbusier. Known as *le Plan Voisin*, it was of course called a piece of futuristic madness and rejected, much to the disadvantage of anyone who tries to drive a motorcar in Paris today.

*   *   *

It has been said that Gabriel Voisin had three passions in his life: aeroplanes, motorcars and women, though probably not in that order. As for the last-mentioned passion, to which a certain amount of space is devoted in the two volumes of Voisin's memoirs, it is characteristic, not of 'being French' or some such nonsense, but of an attitude whose models must be sought in the eighteenth century. I have referred earlier to the eighteenth century in discussing Voisin's talents as a polemicist, and it seems to me that his behaviour is all of a piece in this respect. The scientific outlook of the Age of the Enlightenment, its attempts to base the rules of practical life on reason, its impatience with superstitions, delusions and complacencies of all kinds—all these are typical of Voisin's *Weltanschauung*, and this is equally true of his unselfconscious sensuality.

Calling certain parts of Voisin's writing 'rude', as some have done, is in fact confusing his outlook with lechery and failing to see it as an expression of a civilisation whose values are not identical with ours. What really matters is that this implies a failure to recognise Voisin's outlook as a clue to the context of his other activities, a context in which all the things he did or made fall into place and assume the meaning and purpose for which they were intended, instead of remaining a series of disparate technical creations, outstanding only as such. In other words, Voisin not only made exceptional machines, he also attempted to create the civilisation in which they would be meaningful, a twentieth-century Enlightenment, or, yet again: he tried to find a way in which we, in our machine age, can be civilised *in terms of* our machines, instead of in spite of them.

From this point of view, there is little in common between Voisin and even outstanding engineers such as Lory, Bugatti, Sainturat or Birkigt, whose creations are traditionally compared—justly if perhaps a little tritely—to works of art. It was the very premises of this art that Voisin tried to change, and did change.

*   *   *

Very few of Voisin's creations have been preserved. Of some, there are photographs, of others only descriptions, and there must be quite a few that are already forgotten. When studying this material, I often thought of Niepce de Saint-Victor, who (around 1850) invented an emulsion for photographs in colour, but failed to discover a substance that would fix the images. And I always imagine him—Abel was his name—standing in his laboratory and watching the images appear in the developer, beautiful scenes in stunning colours, but powerless to hold on to them as they vanished again. I might as well admit that I did have an emotion or two when I contemplated the sombre, unlit buildings of the former Voisin works in Issy, reflecting how

> Some burn damp faggots, others may consume
> the entire combustible world in one small room
> as though dried straw, and if we turn about
> the bare chimney is gone black out
> because the work had finished in that flare.

10 ALEC ISSIGONIS

by RONALD BARKER

SIR ALEC ISSIGONIS, CBE (1906–1988). British:
Responsible for the Morris Minor (1948), and
transverse-engined fwd Austin/Morris Mini
(1959), 1100 (1962), 1800 (1964), and Maxi (1969).
Possibly the most colourful and creative design
personality of the post-World-War-II era.

*To bring motorcar travel within the reach of millions is surely to have
contributed in the highest degree to human happiness, human sociability,
and human broadening of the mind and spirit in the last half of our
century. Alec Issigonis, the designer of the Morris Minor 1948, the
Mini-Minor and Austin Seven 1959, the Morris 1100 1962, has
enriched the lives of millions of people in every decade since the war.*

Professor Angus Wilson, University of East Anglia

ALEC ISSIGONIS

ALEXANDER ARNOLD CONSTANTINE ISSIGONIS will never forget the day the grey submarine went down, leaving the pungent oily smell to linger in his nostrils for the rest of his life. It happened about 1912 in the Bay of Smyrna, off the Turkish coast, a disaster observed by only a handful of eye-witnesses and never reported in the world press. He was then a small boy of six, sensitive and impressionable. With his mother, he had been rowed out to meet the tall-funnelled steamship that had just sailed up the gulf from the Aegean, bringing his father home from a visit to an eye specialist in Vienna. They scrambled aboard to find him, as usual, laden with presents, and after the family greetings he handed the boy an exciting long parcel. 'Here, Alec my lad, something for you!' Frantic little fingers worked at the string and ripped the package apart: a submarine! Yes, grey and with the strange oily smell when you sniffed down its conning tower.

Where better than the Bay of Smyrna for proving trials of a new submarine, thought the young Issigonis. So the moment they were back in the tender he wound the clockwork up tight, reached over the boat's side to dip it gently into the water and let it go; for a moment the submarine rested on the surface, then nosed down just as the instructions said it would. He watched as the graceful metal hull distorted into a shimmering grey patchwork before gradually dissolving from view, never to reappear. . . .

Smyrna, his birthplace, is now called Izmir. The population then was predominately Greek, and grandfather Issigonis was a well-to-do Greek national who had established an engineering business, manufacturing mostly hand-pumps for marine and other purposes. He sent his son— Alec's father—to England in his late teens to further his education, and Constantine Issigonis had such a roaring time and so loved the country that he remained there until his father's death in 1900 and became a naturalised subject of Queen Victoria. Returning then to Smyrna to take over the engineering works with his brother, together they developed its activities, making boilers and adding a foundry for casting cylinder blocks for marine engines, as well as undertaking repairs to ships that put into the port with mechanical troubles.

Among the wide cosmopolitan circle of the Issigonis' friends was a Bavarian family that had built Smyrna's first brewery and Constantine, now in his early thirties, courted and married their nineteen-year-old daughter Hulda; it was their only offspring who was destined over half a century later to conceive the Mini, that masterpiece of creative design

which added a new term to international dialogue as well as revolution-
ising small-car design philosophy.

Seen in retrospect they were halcyon, blissful years, when he lived in
a world of Franz Lehar music and the influences of later life were
absorbed joyfully and unconsciously. Young Alec loved to wander around
the marine factory, and in particular to stand and gaze at a big stationary
engine driving all the plant—pulsing rhythmically and hissing steam;
the fascination for steam never deserted him.

He watched an aeroplane fly before ever setting eyes on a motorcar.
One day in 1912 his father took him out to the racecourse at Paradise,
over the hills behind Smyrna, and there from the grandstand they
witnessed the French aviator, Adolphe Pégoud, lift his Bériot monoplane off the ground as if by magic, circle around and bring it safely
back to earth. It was not until the following year that Pégoud, in France,
first demonstrated in public the loop and other aerobatic tricks, which
previously had been mostly unintentional.

Not all the influences were mechanical, however. It was a Greek
custom to give children young lambs as pets and then, at Easter, cruelly
to slaughter them while the children cried their hearts out. Alec begged
for his to be spared, but it grew later into a large and ill-tempered ram
that used to chase him and his young friends around the paddock.
Nevertheless, to his last day he could not bear to eat lamb chops. 'But I
adore red currant jelly!' he would add.

When the war came in 1914, as British subjects the Issigonis family
found themselves trapped in enemy territory. Their first experience of
hostilities was when a British-crewed Farman biplane began flying over
each morning about 6 am from a base on the island of Mytilene to drop
a bomb or two. The observer lobbed them over the side of the fuselage
like footballs, and they did little damage beyond killing the odd stray
goat or donkey. Then the Germans sent a squadron of Fokker fighters
to bring the Farman down, and Alec's nanny took him to see the still
smouldering wreckage on the racecourse. Having escaped unhurt, the
crew had set fire to it. Yet the Turks seemed to bear them no malice
and treated them kindly, even escorting them to church on Sundays—
it was that kind of war to begin with. Alec was also taken out to the
German air base, where friendly guards lifted him into a Fokker's
cockpit. But ultimately the German army arrived in force, interned the
Issigonis family together with the rest of the British colony, and took
over all their property.

After the war, Prime Minister Lloyd George negotiated a treaty
between Greece and Britain, making over to the Greeks a narrow strip
of the Turkish coast including Smyrna. So they had seen the Turkish

army, the German army, and now the Greek army; political shuttle-cocks, they were once again briefly on the winning side of the net.

It was not until after World War I, in the 'Greek' period, that a few cars came to Smyrna. They were mostly Overlands and Model 'T' Fords, but the one that made a lasting impression on the boy Issigonis was a V-8 Cadillac belonging to a Mr Smith, who worked for the American Standard Oil Company. 'I was absolutely enthralled with it and became great friends with the chauffeur, because he would take me for short rides whenever no one was looking. This car ran with no noise at all; he used to drive it with the tyres running on the tramlines, which were just the right gauge, to avoid the rough cobbles. It was then perfection—not a sound. In those days I was not even aware of the existence of the Rolls-Royce.'

In 1922 the Turks came back and threw the Greeks out again. It was full-scale war. Smyrna was set on fire, and the British colony had to leave at a moment's notice in Royal Navy warships that were in the harbour. Alec's father was already in poor health when they were taken to Malta as refugees. Although they made reparation claims, by the time a commission sitting in Paris had sorted things out all the money had gone.

In all the turmoil, trapped in Smyrna with his family as enemy aliens, there had been no question of a formal education for Alec. He had an elderly British governess, an Irish tutor tried without much success to coach him in mathematics, and a Greek woman taught him the rudi-ments of her language, long since forgotten. So he was virtually uned-ucated on conventional lines when, in 1923, his mother took him to London at the age of seventeen and set him up at the Battersea Poly-technic. She gave him ten shillings a week pocket money—very gen-erous in those days—and then went back to Malta, where her husband died soon after her arrival. She then returned to Smyrna to see what could be salvaged from the property; the factory had been destroyed, but she sold the land and came back to London.

At Battersea, Alec failed three times to matriculate, so a BSc was out of the question and the only alternative was to take a diploma course in mechanical engineering; even then, in his final exam, he could not pass in mathematics, a subject he loathed. 'All creative people hate mathe-matics. It's the most uncreative subject you can study, unless you become an Einstein and study it in the abstract, philosophical sense, as to why numbers and things exist.'

When he had just scraped through his exams, his mother bought him a 10hp Singer with an enormous Weymann saloon body. In this car, she said, they would do a Grand Tour of Europe, because he needed educating. Although it was brand new, they had only reached Versailles when the big-ends ran. So while his mother and cousin travelling with them did the rounds of the Palais, Alec was busy helping a garage mechanic to refit bearings and get the car running again. They continued down to the Riviera and 'did' all the main resorts eastward to Monte Carlo—then returned by way of Switzerland. By that time there were no inner tubes left and they were having to stuff the tyres with grass to get from one village to the next. That was the finish of the Grand Tour—the rest of Europe would have to wait for another time. But the Singer went on to serve him for 80,000 miles before giving place to a Gordon England sports Austin Seven.

Alec's first job, in 1928, was in a London design office working on a semi-automatic transmission that was going to turn the world inside out. 'Whenever I go down Victoria Street I always take my hat off at No 66,' he says. This device employed a conventional plate clutch and gearbox with a freewheel added on the back, and an automatic hydraulic servo to withdraw the clutch when the car stopped. Alec was part-draughtsman, part-salesman, and used to go up to Coventry to try selling these things to Rover, Singer, Humber and any other makers who might be interested. They had demonstration cars, and an arrangement with Coventry Climax to use their premises as Midlands headquarters. But this project, and others like it, were doomed to failure when General Motors introduced synchromesh, and Alec soon realised that he was wasting his time.

Through his activities in Coventry he had met a brilliant man named Wishart, then chief engineer with Humber, and was invited to go and work in their drawing-office; that was about 1933. The prospect of being concerned with cars as a whole rather than confined to one component was very attractive to him. As he had always been specially interested in suspension problems, Wishart put him to work in that department under Bill Heynes (later to become engineering chief of SS and Jaguar), and so began another firm friendship.

Another of Heynes' team was a very bright young man called Clapham, also mad about cars, and he and Alec together persuaded the management to let them build a Hillman Minx with independent front suspension. They designed all the parts, and when these had been made and assembled on the Minx, they rushed down to the experimental shop to try it. Heaven forbid—the steering was reversed! Nevertheless, they begged the chief experimental engineer to let them give it a try down the works drive, but very wisely he refused. The geometrical oversight was soon corrected and the ifs installation proved very good, though it never progressed beyond the prototype stage. However, in 1936 Issigonis

Page 295 (*left*) Alec Issigonis in Smyrna at a tender age; (*below*) Sir Alec Issigonis at Longbridge in his 'space age'

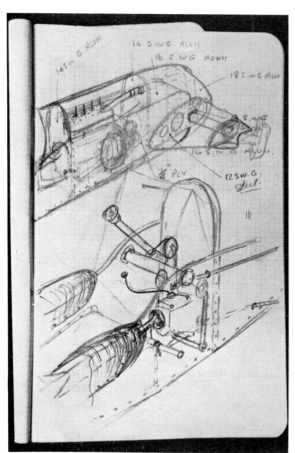

Page 296 (*above*) Sketch-book designs for the Lightweight Special; (*below*) the Lightweight Special at Silverstone with Alec at the wheel. Behind it are co-constructor J. M. P. 'George' Dowson (in glasses with paddock pass), and Charles Griffin (in dark suit), chief collaborator at Longbridge in all Issigonis projects

had the satisfaction of seeing Humber and their associated firm Hillman bring out new six-cylinder models with a transverse leaf ifs to which he had contributed. But he remembers being frustrated at not being allowed to use coil springs because a Rootes subsidiary company made leaf springs.

Working in the Humber drawing-office on these projects, his mind drifted far away at times, dreaming about a very special sprint racer he was about to start designing. The idea had been sparked off by a super-charged Austin Seven 'Ulster' acquired as a spare-time diversion a year or two earlier—at the cost of his mother's precious solitaire ring. With this he took part in week-end sprints, such as the Lewes Speed Trials and Shelsley Walsh Hill Climb, and it was not long before he began trying to iron out some of the car's deficiencies and to develop it further. Ultimately it became a semi-*monoposto* (wide enough for two, but only one seat) with the rather flexible chassis frame stiffened by stressed

'Even-keel' independent front suspension of the Humber Snipe and
Hillman Hawk, 1936

bonnet sides and radiator cowl. His ally in these and other substantial modifications was George Dowson, then working for English Electric at Rugby but sharing lodgings with Alec in Coventry. Murray Jamieson, who concocted the little 'works' Austin racers including the famous 750cc twin-cam, recognised their efforts by helping out with special bits and pieces designed for his racing engines.

In 1933 Mrs Issigonis rented a house in Kenilworth, where the garage was immediately commandeered for construction of the new Lightweight Special. In no time a full-scale arrangement drawing began to take shape on one of the walls. Almost every part had to be made slowly and painstakingly by hand without power tools even for drilling, as and when money could be found to pay for the materials. Meanwhile, the Ulster was progressively devoured piecemeal by the Lightweight, so he had to buy an Austin Heavy Twelve for daily transport—one of the best cars he ever had. The Lightweight was still far from complete when, in 1936, Alec moved to the Morris factory at Cowley, near Oxford, and set up house with his mother in Abingdon; in fact, it did not run until 1939, a few months before the war put an end to such activities.

He subsequently shrugged it off as 'a frivolity in my life. It was not so much a design exercise as a means of teaching me to use my hands. George and I learnt the hard way how to build something for ourselves from scratch.'

Yet it was unique in design and structurally very advanced for the 1930s. A stiff monocoque, it had side members of plywood faced with aluminium sheet and united by steel tube cross-members, the bulkhead and seat pan; the final drive casing and rigidly mounted crankcase were also stressed. Suspension was all-independent; at the front the fully cowled upper wishbones operated bell-cranks which compressed 'springs' contained in a tubular cross-member, these being rubber rings sandwiched between steel discs. The rear wheels were carried on swinging half-axles located by long semi-trailing radius arms triangulating with three tubular transverse links at each side—one below and two above the drive shaft. Rubber loops in tension were the springing media. The cast electron wheels had steel brake liners shrunk into them.

With its original Ulster engine the all-up weight was only 587lb—just over 5.2cwt, and of this the power unit accounted for 2cwt. Soon after the war this was replaced by a small experimental Morris engine with overhead camshaft, also blown, and the power/weight ratio thereupon grew to better than 200bhp/ton. No wonder that for several years its all-too-rare appearances were meteoric—except when it crashed (Dowson driving, on its very first appearance in 1939) or something broke. Getaway traction and road-holding were likewise exceptional,

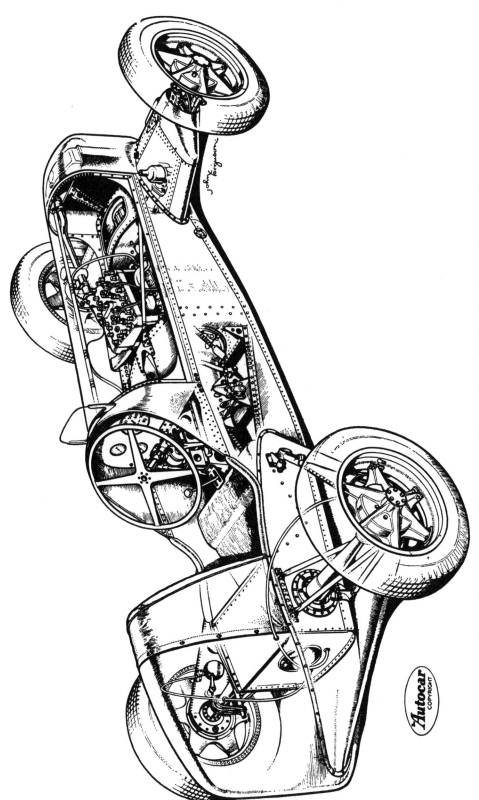

The Lightweight Special in 1939, with supercharged side-valve Austin Seven engine

but not initially, and Alec recalled how he and George chanced upon one of the motoring discoveries of the age. It was at Prescott Hill in 1949, and they were bothered about wheelspin. Removing a rubber loop or two from each side to lower the car gave the wheels negative camber. Their times were lowered dramatically—and today negative camber at the back is a *sine qua non* for competition vehicles.

In the beautifully executed body one detects the influence of German Grand Prix cars of the 1930s—also in the lack of paint and tartan cloth seat covering, both weight-saving moves. For the seat material, incidentally, they went to a tailor and bought a piece of gents' suiting. Running the Lightweight introduced Alec to many pleasant people, racing amateurs like himself, who remained lifelong friends, but if it had any indirect professional value there was no such intent underlying the rubber-suspended monococque 'frivolity'.

Against this spare-time background his professional career continued to progress. He was thirty when he joined Morris, and already was becoming recognised as a young man of above average intellect, with a grasp of engineering matters that promised great things when his talents had matured with experience. Leonard Lord had made Robert Boyle chief engineer, and sent him off to America to study Detroit engineering methods. 'But you must never go to America to learn how to run a business. They do everything the wrong way, yet do it well in the end by just trampling their problems to death. Everything is sectionalised: engine, gearbox, suspension, chassis, body. That way there is no integration, nobody on top to put it all together—except the stylists!'

> 'My old friend Robert came back and told Lord this was the way to do it, so it was decided to start separate departments for engine, gearbox, suspension, chassis—and the back axle, of all things. He fixed up other people to run the first four, and finally was left with me and the back axle. I said: "Not on your bloody life, I won't do a back axle—but I'll do the suspension for you if you like." So the wretched suspension man was switched over to back axles and I was given the suspension.'

Sir Alec freely acknowledged his debt to Maurice Olley, an Englishman who worked for General Motors, for his understanding of automobile suspension and handling behaviour. Probably Olley and Dr Fred Lanchester were the only true authorities on these subjects in the 1930s. Olley originated a lot of theoretical thinking on relationships between spring ratios and dampers, and on the ratio of moment of inertia of the car in pitch divided by its wheelbase. 'Oversteer' and 'understeer' were unknown until he came along to define such things and express them in readily understood terms. 'I believe it was Maurice Olley, then at Vauxhall, who first exploded the traditional theory that the front springs should be harder than the back. Dr Lanchester was possibly the first to understand this, but he was not so explicit and practical; Olley was less academic, passing on his knowledge to the workshop mechanic rather than to some learned institution.' Maurice Olley remained one of Alec's professional heroes.

In no time the separate specialist departments at Morris created by Boyle suffered through lack of cohesion. Then Leonard Lord had a tiff with Lord Nuffield early in 1937, 'retired' temporarily at the age of forty, but was soon back in the fray by joining the rival Austin empire at Longbridge, Birmingham, in 1938. Robert Boyle went back to the Morris engine branch at Coventry, and his departmental scheme was disbanded. Meanwhile Issigonis stayed at Cowley to continue his experiments on suspension. Jack Daniels, who had joined MG in 1933 and transferred to Morris Motors when they took MG under their wing two years later, was appointed to assist him, a close union that continued except for a wartime break almost to the end of their careers.

There they built two or three experimental cars with coil spring and wishbone independent fronts, copying the Americans like mad. But Chrysler at that time had a very good conventional beam axle car that rode almost as well as those with ifs. There was one of these at Cowley, and the management said, understandably, that if American cars could manage so well with beam axles, then the Morris engineers must follow suit. So Issigonis persevered with the beam axle and made the Series M Morris Ten, which came out in 1939. That extraordinary genius H. N. Charles* conceived for it a transverse torsion bar with its ends bent back and linked to little brackets sticking up from the axle, to stop this from twisting under braking; an anti-roll stabiliser and radius arms combined, as found on many cars today. With that device complementing very soft leaf springs, much softer than the back ones, the Series M was revolutionary at a time when almost everyone thought the front springs should be the stiffer. The coil spring and wishbone independent suspension finally evolved between Issigonis and Daniels was to have appeared on a Morris at the 1939 London Show, had the war not intervened; it had to wait until 1947, when the GM 1¼ litre saloon was introduced, and the same design remained in use on the MGB well into the 1970s.

Although the Vauxhall Ten was the first British mass-produced monococque, in 1938, the Series M Morris followed very soon after. 'To

---

* Designer of the R-type MG Midget racer of 1934, with backbone chassis and all-independent suspension.

begin with, monococque structures set frightful little human problems. You would take out an experimental car, on which everything was absolutely perfect, and there would be a maddening hammering noise at the back for which there was no possible reason. It might take days to discover a workman had left his pliers in one of the box sections; with conventional cars having separate body and chassis they would simply have fallen out on the road! Then many strange things were left in them—after a week or two a ham sandwich can make the car smell terrible, but such things happen when you pioneer.'

At that time they were in difficulties with directional stability. The Series M Ten was perfectly stable with a beam axle at the front, but the moment the experimental cars were converted to coil spring ifs they ran into this problem. Nowadays one would correct this by altering the roll stiffness front and rear, but there are limits to this. Their trouble was having too little of the car's weight over the front wheels, because it had been built traditionally with the engine well back from the axle. Issigonis did a lot of work on this, studying Olley and also what Chrysler were doing. Chrysler had brought out a car with an outrageously heavy front end, having about 55 per cent of its all-up weight carried by the front wheels, whereas most cars then had a tail-heavy bias and consequently would not steer properly.

Realising this, he decided that if ever he designed a complete car it must be nose-heavy, although traction would be the poorer for losing weight over the driven rear wheels. Years later the Morris Minor (1948) put this thinking into practice; it was balanced around 55-45, and right from the beginning there were no directional stability problems.

He spent the years of World War II with Morris, making strange devices for the Army which never got into production. This was not as frustrating as it sounds because they were kept working so hard and it was so exciting; after each setback there was always the next project to keep them occupied. They were put to work on projects so fantastic it seemed obvious from the beginning they could never work, like dropping a mechanised wheelbarrow (with Villiers two-stroke engine) from an aeroplane, with a tiny parachute. It was dismantled and packed into a long cylinder, and easily put together again—in theory—after the drop. But when the thing hit the ground it was usually all smashed to pieces. One of their test methods was to hang the cylinders from trees and cut the rope. They spent months dropping wheelbarrows and driving them about in a forest somewhere up north, carrying a 500lb pay load. The idea was that one man could wheel a heavy load through a Burmese jungle, for instance, provided there was some sort of track.

'We also tried out an amphibious version on the lake at Blenheim Palace—still a wheelbarrow carrying a 500lb pay load, and when you came to a river it would float. It was boat-shaped, with oars as well as wheels and tyres. We took one of these down to the Devonshire coast one day to demonstrate to some naval commander, and the first thing they did was to put me in it and tow me out to sea—quite rough sea—to see whether we would sink.'

After D-Day in 1944, when the Allies moved into France and the end of the war looked more imminent than it was in fact, Issigonis began serious design studies for a small car, which had been ticking over at the back of his mind while wartime activities were absorbing every scrap of his time.

By now he was convinced that one couldn't work effectively on a suspension system without a car to stick it on to. It had to be matched to its environment, and a car must be designed as an integrated whole. He suggested this to the people then in charge, who agreed to give him a small facility for the purpose, which is really how the Morris Minor started. Jack Daniels—a brilliant draughtsman—rejoined him to help with detail design, and he had a man called Job to draw the body. 'It's tremendously important when designing a car—if it's to be any good and last for many years—to have only a handful of men working on it, not a great emporium of draughtsmen; that way you get nowhere, communication is impossible.'

Among the engine layouts they contemplated for the new Minor was a side-valve flat four. It was very simple and ran well after a few deficiencies had been sorted out, but almost inevitably lack of finance for tooling quashed it, and they had to adapt the car to take the old vertical four already in production for the Series E Morris Eight.

Issigonis was responsible for the Minor's body, too, and at that time was going through an American phase—all voluptuous curves, front mudguards extended into the door panels, 'vee' windscreens. Initially he made it the same width as the Series E, and it looked ridiculous—far too narrow. 'One night, in despair, I got my mechanics to cut one right down the middle; they moved the two halves apart, and I stood some distance away saying: "No, that's too much, a little closer, closer—stop!" And the difference turned out to be four inches.'

So, whereas the original prototype was 56 inches wide like the Series E, it ended up at 60 inches, and this was done for aesthetic reasons alone, not to improve the roadholding because the narrow cars were handling very well with their heavy front ends. Issigonis was convinced the Minor would never have survived had it been left narrow. 'Proportions are everything. When I study a car to assess its looks I don't say: "It's pretty," or "It's well styled"; I say: "Does it look elegant?" In other words, are the proportions right? The Ancient Greeks knew all about this—the columns in their temples have perfection of proportion. Being Modern and only half Greek, I have inherited no more than an average instinct or flair for aesthetics. But so many people never try to discover their latent talents in this direction, let alone to develop and exploit them.'

The narrow-bodied prototype Morris Minor with side-valve, flat-four engine and three-speed gearbox, 1947–8

About that time, they also experimented with wide-rimmed wheels on the Series M 10hp cars (which were continued post-war for a few years) but ran into harshness problems. The harshness of a tyre is determined by a triangle drawn between the two sides of the rim and the centre of the tyre tread—the bigger that angle, the harsher the tyre becomes. Since those times tyre makers have developed wide-section tyres with softer walls and car makers have learnt how to disguise or absorb harshness in resilient suspension mountings.

He would like to have given the Minor front-drive, having come to appreciate its virtues in pre-war and contemporary Citroëns, but this was quite out of the question through lack of time and facilities. To give the car the speculative nose-heavy bias by allowing room well forward for the flat-four engine originally planned, torsion-bar front suspension was used with the bars running longitudinally and carrying the loads well back into the main structure of the monocoque. He experimented with an independently sprung rear end as well as the

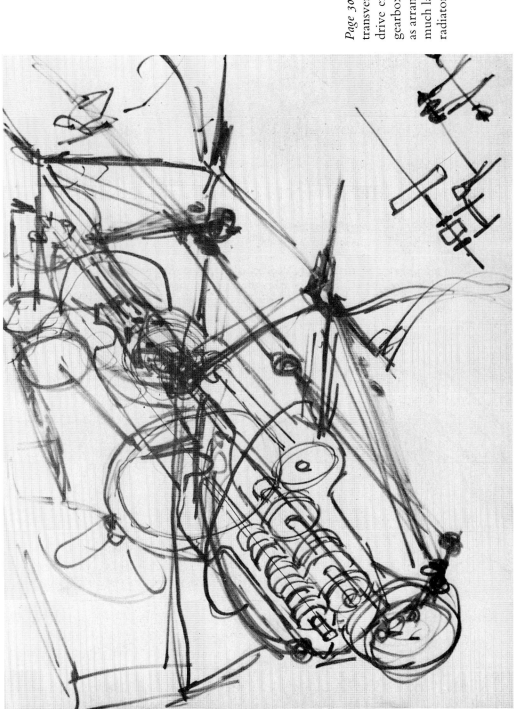

Page 305 First thoughts on transverse engine and front drive envisaged a separate gearbox behind the flywheel, as arranged in Dante Giacosa's much later Fiat 128, and the radiator in the nose

Page 306 (above) Morris Minor 'doodles' during the American phase; (below)
Issigonis taught himself the art of fabricating integral body structures from sheet
metal. This sketch details the Morris Minor rear seat pan and wheel arch

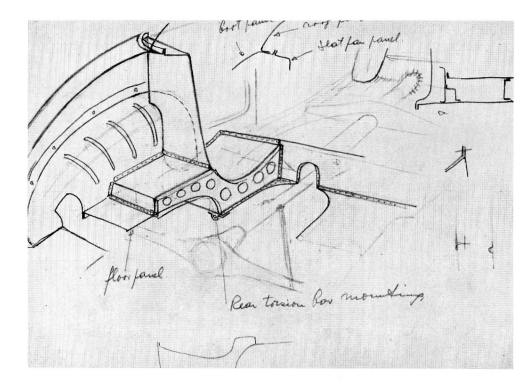

foot pan   cross t

Seat pan panel.

floor panel

Rear torsion bar mounting

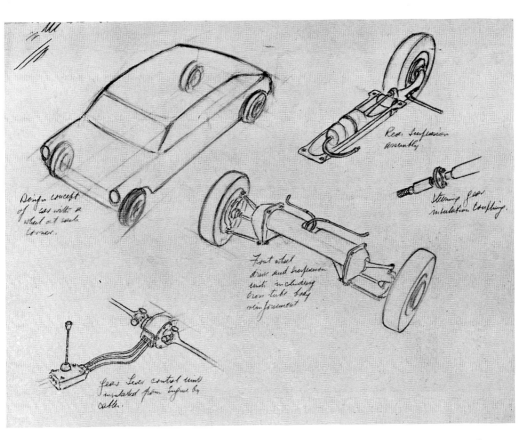

Design concept
of car with a
wheel at each
corner.

Rear Suspension
assembly.

Steering gear
installation coupling.

Front wheel
drive and suspension
units including
cross tube body
reinforcement

Gear Lever control units
insulated from Engine by
cable.

Page 307 (*above*) Front and rear Hydrolastic suspension units and other details projected
for the Austin 1800; (*below*) an early study in torsion-bar suspension allied to transverse
engine and front drive

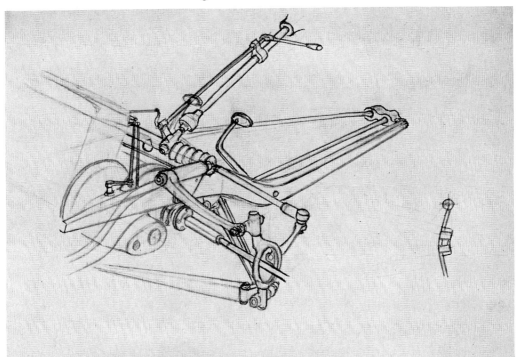

Page 308 (above) The first Issigonis miniature was this wartime amphibian, intended for delivery by parachute in jungle areas; (below) by leaps and bounds the Mini gained prodigious success in international rallies. This is Finnish driver, Timo Makinen, on a training 'flight'

combination of a live axle with torsion bar springs, but there was far too much to be done and time was too short to develop either of these.

The Minor was probably the first mass-produced car to have the headlamps embodied in the grille, but it had been in production only a few months when an American regulation stipulated a minimum height above ground level for headlamps, and there were rumours that this requirement would spread all over Europe. Raising them meant bringing them out into the wing pressings and spoiling the car's lines, but there was no alternative.

In planning a new small car he had gone completely against the grain, for the general reaction to wartime privations and restrictions was: 'To hell now with petrol-rationing and tight-fisted economy—let's have bigger and more powerful cars!' But Issigonis guessed rightly that petrol rationing would not be dropped quickly, knowing well enough that the end of a war did not herald immediate universal prosperity for the victors. Money was devalued, and manufacturing costs had risen sharply since 1939. There was bound to be a huge and hungry market for a small, up-to-date four-seater, cheap to buy and cheap to run.

Meanwhile, most of the rival firms compromised by hurriedly reintroducing small cars of pre-war design while planning larger-engined ones to replace them. Fortunately the Morris management fully supported Sir Alec's intuition and deductions and were solidly behind him in the Minor project, otherwise he could never have gone ahead with it. Were they right? A twenty-four-year production run and well over 1½ million off the lines at close of production in 1972 fully justified their faith in him.

Apart from the repositioned headlamps, curved single-panel screen and larger rear window, the Minor's looks remained practically unchanged, but it was revitalised twice mechanically. In mid-1952, following the Nuffield-Austin merger, the 918cc side-valve Morris engine (27bhp at 4,400rpm) was replaced by the smaller but more efficient ohv Austin A30 unit (803cc, 30bhp at 4,800rpm), necessitating a lower final drive ratio which shortened its legs but improved acceleration. In 1957 it was upgraded with a 37bhp 948cc engine and renamed the Minor 1000, and subsequently the process was repeated with a 50bhp 1100cc unit without a change of name.

Fearing that the merger would breed political squabbles, and feeling very strongly that politics and engineering are bad mixers, Sir Alec decided it was time to get out and moved to Alvis, where he stayed two or three years and worked extremely hard designing a car completely new from stem to stern, including the engine. It was to be a sporting family car, exceptionally compact and light (about 22cwt), not much over 13ft long but very wide and a full five-seater, powered by an all-aluminium 3½ litre V-8 engine with single overhead camshafts. An ingenious combination of a two-speed gearbox and electrically controlled overdrive in unit with the final drive gave it four speeds, and it had all-independent rubber suspension under Moulton patents. Towards the end of the project they began experimenting with hydraulic front-to-rear interconnection using one common spring for all four wheels. Although the engine was not highly tuned the Alvis V-8 would do about 110mph, and they got as far as doing full-scale tests with the prototype, Issigonis and two of his draughtsmen taking turns at the wheel during some high-speed endurance runs around MIRA's test circuit near Nuneaton to cover 1,000 miles a night for a week.

> 'It was completely out of character with Alvis tradition, but I still think the concept was good and John Parkes, the company's chief, was very much behind us. However, when we came to consider the cost of tooling the body it was out of the question for a firm of that size. There were so many problems, perhaps it was just as well it never happened, and in a way I was relieved when Sir Leonard Lord rang me one day in late 1955 and said: "Come back to Longbridge!" So early in 1956 I went back to Longbridge, and Alvis meanwhile wisely destroyed the car and all the drawings. I was glad to be back with Lord—a tough, wonderful man with a fantastic personality, a born businessman and a great production engineer.'

Issigonis brought with him from Alvis two first-class men, Chris Kingham and John Shepherd, and Jack Daniels naturally gravitated back into the fold. Their first joint effort was a front-engined (single ohc 1½ litre) rear-drive saloon having all-independent rubber suspension with hydraulic interconnection, a further development of the system tried on the Alvis. The prototype looked very similar to the front-drive 1100 that was to follow years later, in 1962.

Then came the Suez crisis in September 1956, when Egypt's dictator Nasser suddenly closed the Canal and his Arab confederates in Syria cut the main oil pipeline crossing that country. The resultant petrol shortage, which looked like it would continue indefinitely, brought about a rash of 'bubble-cars', three- and four-wheeled economy runabouts powered by motorcycle engines and encased in egg-like shells with moulded plastic tops. 'One day Sir Leonard said to me: "God damn these bloody awful bubble-cars. We must drive them out of the streets by designing a proper miniature car"—and that was how the Mini started.'

It was in March of 1957 that the decision was taken to put aside the 1½ litre project and concentrate all their efforts on the miniature. Starting absolutely from scratch, it had to be designed, developed, tested and tooled up for production in the shortest time possible. This task was the more welcome because Sir Alec always preferred small things and was unmoved by mammoth objects such as skyscrapers, jumbo-jets or the QE2. The thought of Texas left him cold, and when American friends once drove him against his will to see the Niagara Falls he chose to stay in the car and read a book. 'I was weaned on Austin Sevens when all my friends had Bentleys and Vauxhall 30/98s, and I am still not much interested in large cars. A small one sets a tremendous design challenge non-existent with a large one. Mr Royce had nothing to do!'

Like all Issigonis designs, it first began to take shape in freehand sketches, rather than as a list of formal requirements and technical formulae. His now well-known talent for freehand sketching of engineering subjects had this in common with Leonardo da Vinci, that what elegance of form some of them may have had is coincidental, in the sense that they were produced simply as a practical means of transmitting lines of thought or guidance, without any artistic pretensions. He asserted that unless one could sketch or draw one's ideas one could never communicate any engineering thought to anybody. 'I forced myself to develop this particular aptitude simply to help myself to communicate technically with my associates. One acquires a style of sorts, if you like, but inherently it was only done to make my work possible.'

An interest in drawing and skill as a formal draughtsman may have been inherited from his father, and the hours spent watching him at work on the drawing-board always kept in his dressing-room. 'When we used to go and stay with relatives at their farm near Ephesus, the most precious things I could have were a pencil and a copybook full of clean pages—for paper was scarce in Turkey then and quite expensive.'

Whether considering a new broad concept or puzzling over a solution to some elusive detail, he would 'talk to himself' through doodles and sketches, and when discussing ideas and schemes with colleagues his fibre pen or soft pencil was scarcely still for a moment. He had a strong penchant for Monte Carlo, and over the years took to escaping there occasionally for a few days, to contemplate and scheme in peace and sunshine on the point by the Beach Hotel, or in the *belle époque* splendours of *l'Hermitage*. Old notebooks have survived him with sketches of the Lightweight Special sprint car when it was just a twinkle in his eye.

Once, when I went to see him in his Longbridge office during his term as overall engineering chief of the British Motor Corporation, there was a sheet of unlined foolscap on his desk with a vertical row of perhaps twenty thumbnail sketches of vehicles, private and commercial, plus several outlines of engines. 'Oh,' he said, 'before you came in I was making a list of all the projects I have in hand and putting them into some order of priority.'

The only restriction in the design of the new miniature was that, as there was no time for developing and tooling a new engine, this had to be based on the existing four-cylinder water-cooled A Series. We have already seen how Sir Alec's earlier work had shown him stability depended on nose heaviness, so the engine must be at the front. For good traction a very small car needs to have as much weight as possible concentrated over the driven wheels, so with the engine forward it must have front-wheel-drive, and in terms of space the most economical way of accommodating an in-line four with its transmission would be to set it transversely. Sir Alec had thought of this idea before; in fact, on the wall of his dining-room was a framed sketch he did about 1950 for a transverse-engine front-drive adaptation of a Morris Minor, arranged very like the Fiat 128 with engine, clutch and gearbox on the same plane and the final drive offset below. It differed, though, in having equal length outer drive shafts, with an intermediate jack-shaft between them. He and Daniels built a front-drive Minor in this form, but it was completed only just before he departed to join Alvis in 1952 so he had scarcely a chance to drive it.

> 'It was abandoned under a dust sheet until one day the following winter, when my old colleagues still at Morris took it out to try it round the test ground on ice and snow, and were absolutely amazed. I didn't know all this until I rejoined Austin, when Jack told me how wonderful this front-drive was. I said to him: "We must have front-drive on the new car."'

But where could they find constant velocity joints for the front drive shafts? While working on the fwd Morris Minor they had tried and rejected a constant-velocity system using two Hooke's joints inside the wheel hub, so that the king-pin centre line passed between them to divide the shaft in two, in effect. The main snag was that the hub bearing had to be almost as big as a brake drum, and appallingly heavy. The solution arrived almost casually. 'One day I was busy sketching something in the Kremlin* when Sid Enever of MG sent in a note with a drawing of the Hardy-Spicer Birfield joint: "You might be interested in this," it read. So I rang Hardy-Spicer and asked what they were using it on. Apparently it was in production for some submarine control gear. But for Sid's note and the Birfield joint there might never have been a Mini as we know it.'

It's one thing to have a brainwave, quite another to have the self-assurance and courage to go through with it. The solution he hit upon to abbreviate the engine-clutch-gearbox unit was to place the gearshafts in the crankcase, directly beneath the crankshaft, and unite them through

---

* 'Kremlin': Works sobriquet for the Austin headquarters building at Longbridge.

idler gears; moreover, to leave the clutch where it was most accessible (being a consumable item) at the back of the crankshaft, its driven plate splined to a primary drive sleeve running on the crank's tailshaft to take the motion back into the crankcase. However confident Issigonis and his associates might have been about the long-term wisdom of having engine and transmission 'share the same bathwater'—a common oil supply—some sceptics predictably raised a hue and cry about this when the model was released.

With this enigma resolved at least in principle, Alec could proceed with planning a body, 10ft bumper to bumper, of which no less than 80 per cent could be given over to passengers and their baggage, certainly unprecedented among front-engined cars. Mainly to keep the wheel arches down to minimum dimensions in his quest for passenger space, but also to reduce unsprung weight and balance the car's aesthetic proportions, Issigonis enlisted the co-operation of Dunlop in making special tyres to fit 10 inch wheels—roller-coaster size by comparison with what other car makers were fitting. The ADO 15 (Austin Design Office project 15) was also the first mass-produced car to have rubber springs for its all-independent suspension, a joint development by Alex Moulton, the Issigonis team at Longbridge and Dunlop, who made the moulded rubber elements.

In simple terms a fat circular rubber diaphragm reduced to minimal thickness in the centre is compressed between two cones, one attached to the frame, the other linked to a suspension arm. This provides a steeply progressive rate, essential for a vehicle with such wide variations in load between the extremes of driver alone and four up with luggage. Moreover, rubber has inherent self-damping qualities (hysteresis) which reduce its dependence on dampers, and is light and compact in relation to its load-carrying properties.

A brief chronology of the Mini's gestation period may interest readers. Only four months from Sir Leonard's 'go-ahead', wooden mock-ups of the body and principal mechanical parts were completed, and by October 1957 two prototypes were on the road. In July the following year, after a brief run in the car, he told Issigonis to get it on the production lines within twelve months, and the Mini was announced to the public within that period, in August 1959—a most remarkable achievement considering the amount of original thought in the concept and the untried principles and components involved.

It was christened Austin Seven or Morris Mini-Minor, both versions identical but for the intake facade. During the development programme only two modifications were found necessary. First, the engine was switched through 180 degrees to bring the induction system adjacent to the bulkhead, to overcome carburettor freezing problems (and conse-quently expose the ignition distributor to damp and cold behind the grille, which took a little time to sort out); secondly, to mount the complete engine-transmission assembly, as well as the rear suspension, on independent sub-frames. This was necessary to simplify assembly on the production lines and to insulate it to some extent from transmitted mechanical and road noise. The Mini's phenomenal handling properties did not just happen, and in particular Charles Griffin, chief development engineer and himself an expert driver, spent many weeks theorising and experimenting with different geometrics front and rear. One outcome was that the back wheels were given a marked toe-in to reduce the natural oversteer (or more exactly, reduced understeer) characteristic when the throttle is suddenly cut in the middle of a corner.

One of Sir Alec's laconic aphorisms ran: 'Styling is designing for obsolescence.' The Mini was not styled; rather, its contours were evolved logically to suit the purely functional purpose of the car—wheels at the corners, with no 'stylist's decorative overhang', the simplest and smallest shell practical to house four people and the prime mover, something cheap to make yet inherently strong. The end product is not beautiful, but it has an indefinable correctness and harmony of form and proportion that has made it virtually evergreen and classless. Beneath the skin, so to speak, lies a tremendous amount of research, stress analysis, fatigue testing and so on, always bearing in mind the requirements of the production line. Of this, Sir Alec has said: 'In the development of this car we have done far more towards understanding about structural design of bodies than towards understanding about engines.'

If this prodigy at its introduction stirred up more excitement inter-nationally than any previous new model, inevitably it had to wait a while for full public acceptance of a concept almost totally new in dimension, form and mechanical content. Sir Alec denied that he had any thoughts about competition use when laying out the Mini, but he and his associates did demand certain standards of road behaviour to make it exceptionally safe and controllable for ordinary people. Once the various tuning establishments had sampled its qualities, they lost no time with go-faster modifications and Minis were soon in the thick of competition, both track races and road rallies.

It took the BMC three years to jump on this band-waggon commer-cially with the 85mph 997cc Mini Cooper, fitted with disc front brakes to match its performance: thereafter the car's success became astronom-ical. As a portent of what could be done to this car for road use, in December 1962 *The Autocar* tested a 100mph-plus Cooper developed by Daniel Richmond of Downton Engineering. With the engine overbored to bring it to 1,088cc, it averaged 103.5mph with a best run at 106, and accelerated from rest to 60 in 9.6 sec, to 80 in 17.3, compared with 18 and 50.6 sec for the standard Cooper. Understandably, Richmond's close technical association with the Issigonis team began a few weeks later and continued throughout the Mini's halcyon days of competitive stardom, and included development work on its larger sisters.

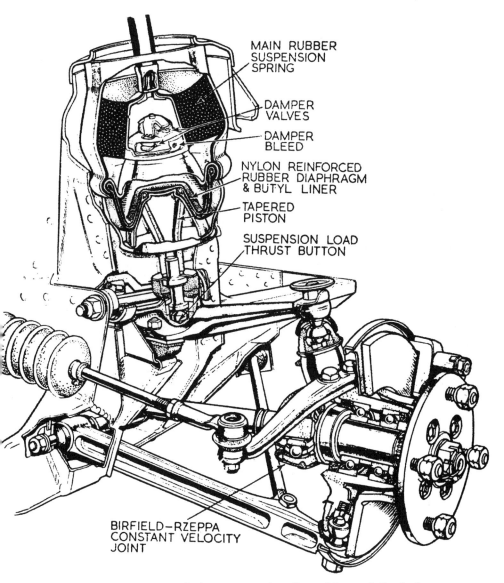

MAIN RUBBER SUSPENSION SPRING

DAMPER VALVES

DAMPER BLEED

NYLON REINFORCED RUBBER DIAPHRAGM & BUTYL LINER

TAPERED PISTON

SUSPENSION LOAD THRUST BUTTON

BIRFIELD—RZEPPA CONSTANT VELOCITY JOINT

Austin-Morris 1100 Hydrolastic suspension, front drive and disc brakes

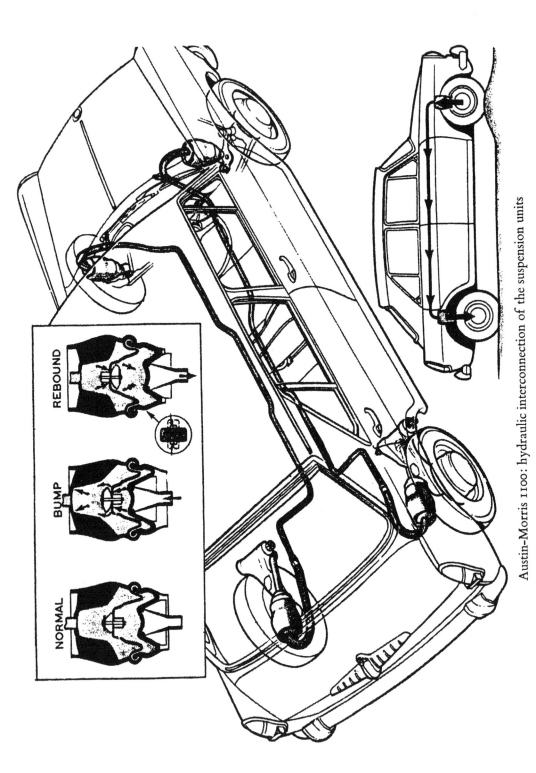

Austin–Morris 1100: hydraulic interconnection of the suspension units

Page 317 (above) The Morris Minor at the time of its introduction, disliked by Lord Nuffield, but destined to become, along with the Mini, Issigonis' most enduring design. (Photo courtesy of British Motor Industry Heritage Trust/Rover Group); (below) September 1959: All land routes into war-ravaged Algeria are closed, so the author and colleague Peter Riviera rent a Tunisair DC3 to carry their Mini over the border during a strenuous round-the-Med publicity run.

Page 318 (above) Styled by Pininfarina, the Austin-Morris 1100 best-seller, introduced in 1962, used a compressed gas suspension system with fore-and-aft interconnection. (Photo courtesy of British Motor Industry Heritage Trust/Rover Group); (below) 1968: Austin's challenger to France's innovative front-wheel-drive Renault 16. The hatch-back Maxi was roomy but stolid, lacking the Renault's flair and style. (Photo courtesy of British Motor Industry Heritage Trust/Rover Group)

Page 319 Sir Alec Issigonis celebrating his retirement in 1971 amidst his brainchildren. (Photo courtesy of British Motor Industry Heritage Trust/Rover Group)

Page 320 Rather unkindly dubbed the *Great Land Crab*, the 1800 was exceptionally roomy and very strong. It remained in production from 1964 to 1974. From 1969 a 2.2 litre transverse six was squeezed in as an alternative for extra punch and refinement. Total production of both types ran to only 314457. (Photo courtesy of British Motor Industry Heritage Trust/Rover Group)

Richmond's association was abruptly—and ungraciously—terminated by Stokes, presumably with connivance from Issigonis' unappreciative successor, a Mr Webster. Next move was the supplementary 1,071cc Cooper S, with considerably re-worked engine and running gear, including wide-rim wheels and beefed-up brake discs with servo assistance. Later, 970 and 1,275cc versions were added, with net outputs ranging from 65 to 75bhp. Of all the outright victories in international races and rallies, one remembers particularly the Monte Carlo Rallies of 1964, 1965 and 1967, the Alpine Rallies of 1963 and 1967, the Tulip, German and Midnight Sun Rallies in 1962, the RAC in 1965 and Acropolis in 1967. In 1965, the Finnish driver Rauno Aaltonen won the European Rally Championship with a Mini, and on racing circuits the little car took the British Saloon Car Championship three times—in 1961, 1962 and 1969. Other track successes included the European Touring Car Championship in two categories during 1968.

For ordinary citizens, a station wagon came on the market in 1960, and Wolseley and Riley variants followed. The two big interim engineering developments were hydrolastic suspension with fore-and-aft interconnection, and a remarkable automatic four-speed transmission with torque converter incorporated in the crankcase, announced in the autumn of 1965. For 1970 the Mini range was contracted, the standard Cooper, Wolseley and Riley were dropped and 'styling' added a Cyrano de Bergerac nose for new Clubman and GT versions, but from that time its progenitor was permitted no further influence over its development, which remained sadly sterile.

The transverse-engine front-drive doctrine was applied subsequently to the 1100 (1962) which became top-selling model on the home market, to the big 1800 (1965) and the 1½ litre Maxi (1969), all carrying forward the Mini concept of maximum passenger space within modest exterior dimensions, together with well above average stability and dynamic security.

Although Sir Alec designed a number of engines, including the V-8 of the abortive Alvis, none ever reached the production stage. However, his interest and involvement in this pursuit behind the scenes deepened with the years. His view was that the cheap, bread-and-butter piston engine has a long way to go yet; that the future for small cars, including Mini-size, lay in the vertical six-in-line down to one litre capacity tuned to give about 60bhp per litre.

'If only the marketing world would see it, I am convinced this is the next development. I have done enough work to know that a small straight-six can be absolutely fabulous, as quiet and unobtrusive as a V-12. Market research is still hanging on this ridiculous prejudice against small sixes going back over thirty-five years; people who don't understand these things say: "Ah well, they were no good in those days!". But in those days we knew practically nothing about distribution or torsional vibration or bearing materials, for instance. Market research is all bunk!

'We're looking for refinement, not power, and in this respect the difference between the four and the six is dramatic. The small car of today is screaming for refinement, less road noise as well as less mechanical noise, because it's a complete waste of time to have one without the other. Now we could build a Mini as quiet as a big car, and its utility and market penetration would increase because people who have to do hundreds of miles in a day would buy it, as well as those who want it mostly for shopping and short journeys. I'm dedicated to this philosophy. Already the Continentals have solved the mysteries of road noise, but they have not been clever enough yet to make a very small refined engine like a straight-six. Two extra cylinders need not cost much more and the manifolding is simple: everything else can be the same—carburettor, starter, clutch, gearbox and so on.'

From his viewpoint as a confirmed front-drive, transverse engine proponent he favoured the vertical engine as best suited to this layout, because it took the least room. While a compact V-type contributed to passenger space when placed lengthwise in a rear-wheel-drive family saloon, from all other aspects he would choose six-in-line up to about 3 litres and jump straight to a V-12 for larger capacities.

'The trouble with designing a new engine today is that it involves a major management decision on the money it costs to make. In round figures, setting up an engine now costs something like a million pounds per inch of its cylinder block, you see, so someone has to decide where to draw the line. If one could reduce the tooling charges of a modern engine by a half, which I believe is possible, we should not be tied to old designs for far too long, as is now the case. There is no value analysis in plant design, and the problem will remain until the machine-tool designers and planning engineers are made to consider the cost of the plant like car designers have to consider the cost of the car, reducing its price without reducing its quality. If there were less needless extravagance in those quarters, one could have new engines more frequently and engine development would be speeded up. One works away at some entirely new car and all too often the management is forced to say: "Use any engine you like so long as it comes off the current production line."'

By the time Sir Alec had reached the prescribed age of retirement, 65, still bristling with invention and reasonably fit physically, the main components of the UK motor industry were already programmed into a state of near-chaos from which there would be no long-term recovery. As we now know, the USA, France and Japan would eventually pick through the shameful remnants resulting from years of misdirection and militant trade unionism. The decay first came seriously to a head in 1961, when the Leyland truck people suddenly gathered up Standard-Triumph. Next Rover, in 1965, absorbed Alvis, which soon vanished as a consequence, and then in 1966 Jaguar, which had already mopped up Daimler, Coventry Climax and Guy commercials, amalgamated with the British Motor Corporation (Austin, Morris, etc.) together with the Pressed Steel body people to form BMH—the British Motor Holdings Group.

Creative talent rarely goes hand-in-glove with administrative ability, so there was potential for disaster in the combination of the egocentric Issigonis, promoted to technical director, with the gentlemanly but malleable Sir George Harriman as company chief, of an outfit constantly harassed by a rebellious workforce. So the Leyland predator, represented by the ambitious Donald Stokes, sat on the fence watching the BMH prey decline, ready to pounce when the time seemed ripe. The year was 1968.

Predictably, Stokes' henchmen at Triumph were allocated all the key posts, Issigonis the sensitive intellectual being displaced by rough and tough Harry Webster. Antipathy, both social and professional, divided them like a Berlin wall. Hence, while Issigonis was to remain at Longbridge until retirement in 1971, and stayed on the books for a few years thereafter as a part-time consultant, no further manifestation of his talent would ever be accepted or acknowledged by his successors.

In fact, some of the horrors that were to emerge from the Longbridge and Cowley factories represented the very antithesis of his engineering principles. He must have seen the Morris Marina, for instance, as a professional snub in the worst taste. Just as advertising material for the VW Beetle sometimes sought to make a virtue of its known vices in roadholding and handling, so the public was assailed with publicity material extolling the Marina as *Beauty with Brains behind it!* There's a story that a set of snaps taken during a PR shoot in Spain, showing a Marina in the foreground of a flock of sheep, were withdrawn just in time . . .

As a finale to the Issigonis era, already well advanced when Stokes stepped in, the Austin Maxi (1969) really underscored the circumstances that led to takeover. Although following France's clever and very successful Renault 16 with the lift-up one-piece tailgate on a six-window saloon, the concept and packaging were first-class, and on paper the five-speed transmission was a welcome innovation for a front-drive car; the body style, however, lacked both daring and chic. Issigonis once confessed to me that all his designs had seemed to reach the production lines in an immature state, and the utilitarian Maxi was no exception. On arrival it was a mechanical sloth, its lethargy aggravated by over-gearing and an appalling cable-operated shift—and the body rattled! Although improvements soon followed an adverse reception, the marketplace never quite recovers from a bad start.

During this 'lost' period Sir Alec became deeply involved in research into the prospects for a steam-powered Mini. Although he entertained no serious expectations that steam could ever be developed to rival internal combustion for fuel efficiency, he took it as a duty of large organisations to explore every avenue. He was fully aware of other problems: 'Imagine you have left your steam Mini outside London Airport for a week in mid-winter, and everything's frozen up—how will you get home?' However, one day I was invited up to Longbridge and actually witnessed this engine running on a test rig, and beside it stood a Mini specially prepared for the metamorphosis.

Sir Alec's greatest disappointment in those later years was that none of his schemes for fundamental improvement of the Mini were ever adopted. In particular, he wanted to refine it with sweeter engines (including an 1100cc six-cylinder) and a softer ride, substituting simple coil springs for the expensive rubber-based units and at the same time saving weight by casting out the sub-frames. The rubber suspension had been the best possible in 1959, but soon afterwards Renault had shown what could be done with steel for much less money. From time to time I was lucky enough to drive experimental Minis with inches of extra leg room in front, with quiet little aluminium ohc engines and big-car suspension characteristics.

A favourite project was his gearless Mini, externally identical to the standard product, but having a long-stroke, big-torque engine driving through a fixed-pitch hydraulic torque converter. Its only gears were for selecting forward or reverse. The objective was a low-cost urban runabout with basic controls, but sufficient urge to hold its own in any traffic stream. In fact, the carefully tailored engine/transmission characteristics together with drastic weight-saving resulted in a lively take-off and—he had the test figures to prove it—acceptable fuel economy. In retrospect it's such a shame that a genuine Issigonis engine never progressed beyond the prototype stage, so we shall never know what we've missed!

Issigonis was fortunate to become world-famous in his lifetime, and revelled in his celebrity and the social scene it opened to him. He became a cult figure in the '60s when mini-this and mini-that acquired inter-national usage. His engineering abilities were complemented by that mellifluous and unforgettable name, by a sparkling wit and a stage comic's timing as a raconteur. He could have a dinner party in fits over inconsequential topics such as his loathing of lamb chops (". . . but I *adore* red currant jelly!").

National recognition came first with the award in 1964 of a CBE (Commander of the British Empire), and in 1969 he received his knighthood. He also greatly valued his election as a Fellow of the Royal Society and its award of the Leverhulme Medal for 'Services to Science and its Application'. He died in October 1988 in his 82nd year, an old man confined to his Birmingham apartment by ill health—and lonely, because he had cynically cast aside all but a handful of those friends who tried to keep in touch when he was no longer news, but history.

Throughout his professional career and even into retirement, Issigonis made his home with his widowed mother who, incidentally, had wished him to adopt an academic rather than pursue the engineering strain of his father and grandfather. As she became older and more decrepit, their domestic roles gradually changed, and concern for her well-being prejudiced both his social and professional activities, which in earlier years had overlapped to some extent due to his extrovert nature and considerable charm.

It is a tantalising conundrum—whether his creative abilities (some would have it genius) were curbed by the restrictions and shallowness of his home life; and it was profoundly sad, when visiting him towards the end of his days, to witness those talents reduced to pencilling solutions in a book of simple crosswords.

If Sir Alec wasn't the first to combine a transverse engine with front-drive, his Mini package totally revolutionised the concept of the ultra-small car, and introduced a tactile enjoyment and a standard of control to its handling that later developments have never diminished. In 1991 the Morris Minor he created in 1948 is still on the active list, being manufactured in Sri Lanka, and the Series 111 Morris Oxford of the '50s is still the staple diet of Indian motorists as the Hindustan Ambassador. And the Mini, introduced in 1959? Today, in '91, its popularity abroad exceeds that of any other individual British model, and over the years the exported Mini has earned more cash than any other. Young Japanese, too, have adopted it recently as a 'cult' machine in an extraordinary resurgence, and an ancient 'Works' rallying example has fetched £42,000 at auction. It's a story without an end in sight.

11 DANTE GIACOSA

by LJK SETRIGHT

DANTE GIACOSA (Born 1905). Italian: Aero-engine designer for Fiat who, at 28, designed the mechanics of the miniature Fiat 500 and continued to exercise a commanding influence on European (and World) small car design, culminating in the brilliant front-wheel drive Fiat 128.

DANTE GIACOSA

PHILANTHROPIC ECONOMY is seldom encountered in the motor industry. Economy for the sake of profit, yes; economy for the sake of survival, often; but economy for the sake of the customer, whose needs are felt with empathy and whose wants are seen with sympathy, is rare among those who create cars, for other motives are most often uppermost in their minds.

Henry Ford, although he created the original 'people's car' in the refreshing Model T, was more an opportunist than an idealist; Herbert Austin, catering to all that was smug and self-satisfied in lower-middle-class England, came closer just once when he made the original Austin Seven. Neumeyer (of NSU), Hitler, Porsche and sundry others who collectively formed the idea of the Volkswagen, were variously inspired by greed or—which is much the same—politics.

Only in Fiat, a company formed in 1899 by gentlemen with philanthropic consciences and guided thereafter by men sensitive to the pulse of the Italian people, could a humanitarian designer flourish, devoting himself simultaneously to the discipline of corporate need and the rigour of public want. Of all the great engineers who have had charge of that company's car designs in its long history, the greatest and most influential—and at the same time the gentlest and most sympathetic—was Dante Giacosa, who served them and the Italian people from 1928 to 1970.

His is not a story amenable to that simple straightforward chronology so attractive to the simple reader and the straightforward biographer. Yet, in remarking the lack of chronological continuity in any fully representative Giacosa story, we hit upon the very thing which distinguishes an examination of his work from those accounts of other celebrated designers who merely had careers.

Time and again he had to defer using his ideas and pursuing his principles until many years after the desire had grown within him. There is no synchronicity of thought and time and work in his life, nor even any rational order of events. Other designers in other places did what they could when they thought of it; Giacosa did what he had thought of when he could.

He was born in 1905, and came to Fiat with an engineering degree from the polytechnic in Torino. He did not come directly, although he had spent five years at his studies: Graduating in 1927, he then had to present himself to the Officer Training School at Bra for his military service. At the end of the course in June 1928 he was graded second out of all the students involved—a better result, he later reflected with some amusement, than he had secured at the Turin Polytechnic.

He might well have been proud, as well as amused: He was never in any doubt, throughout his working life, that discipline was a vital tool in the design engineer's kit, and that the discipline had to be self-imposed, to come from within, whether by nature or by training. The nature was there in him already, nurtured at an early age by the modest circumstances of his parents and family in which economy, thrift, and the intellectual challenge of making things last received daily emphasis. At the Polytechnic he pursued his aversion to extravagance (which he would have seen as literally wandering off course), ignoring fashionable diversions, and especially ignoring politics.

All his life he would carefully and conscientiously avoid politics, despite the political turmoil which filled Italy as Mussolini came to power in the early 1920s, despite the political repercussions of the nation's involvement in the wars of the 1930s and 1940s. Despite being at the eye of many an administrative storm in the factory, he likewise avoided office politics, at every level. He felt that they were none of his business, which was to get his head down and do to the best of his abilities the work for which he had been engaged. Whether as a new draughtsman or as 'chief engineer'—Director of the Motor Vehicle Technical Offices, as he became in 1946—that was what he did, and the respect that he earned by his conduct rivalled that earned by his work.

Could it have been that final year of military training which finally polished his skills in handling people? He described discipline as a spontaneous emanation of culture and upbringing, from which he derived not only a stringent self-control but also a strict respect for the hierarchy. Fellow workers, superiors, subordinates, people from outside his organisation, all found him a model of cool reason, graceful manner, and firm principle. He was to state, with quiet pride after he retired, that he had never given an order; what he always asked for were collaboration and understanding, which were exactly what he also sought to give.

Such a man, less susceptible to corruption than the majority of men scaling any typical industrial pyramid, acquires authority with grace. In his last decade of service to Fiat, he was still the mild-mannered man of sensitive mien, his quiet reserve betrayed only by the twinkling of enthusiasm in his eyes when he spoke of some cherished project. Long in the leg and light of build, he had about him an aura that was almost ascetic yet never mean, a spareness that almost deliberately exposed him to the feelings (expressed or detected) of others. He loved technical novelty and engineering adventure, but never let himself be seduced by either; it was always an essentially human concern which pervaded his engineering logic and his professional pride.

It was no less to his credit than to Fiat's that the company cultivated him as carefully as it did. He began work by dealing with a practical and (one suspects) carefully orchestrated score of what seemed like odd jobs. In effect they not only allowed him to play himself in, but also introduced him to a couple of good strong ideas then current at Fiat. One was that competence as a design draughtsman meant a good deal more than a degree in engineering; and that was a common enough attitude in factories all over the world. The other was almost specific to Fiat, the feeling that a good engineer should fear nothing but ought to be as ready and able to design, say, an aeroplane or a submarine as to design a car or a military tractor-train.

Young Giacosa was introduced to a fair smattering of all these within a short time of joining Fiat, and was even given promotion and a salary increase six months after joining. In a couple of years time he would be suffering cuts in pay, as did everybody else, because times in Italy (as in all the slump-ridden world) were hard; but, for the record, he started at 845 lire a month. That was in 1928, the year when exhaustion forced the retirement of the technical director Carlo Cavalli, under whom Fiat had become masters and mentors of advanced engineering technology. It may not be unfair to suggest that Giacosa reached where he did by standing on Cavalli's shoulders; if so, it was because he had the sense of balance, no less than because Cavalli provided the footing.

It is a commonplace observation that no car is ever designed in its entirety by just one man. Few designers have been capable of it, and very few of them had the time for it. Behind every great designer stands at least one man who understands him and can fill in the gaps he leaves: Porsche had Rabe to make his ideas work, Ford had Wills, Royce had Rowledge, and so on. More often, the great man is backed by a full team of design engineers and draughtsmen; this was no less true of Ettore Bugatti than of, say, Charles 'Chuck' Chapman of Opel or Rudolf Uhlenhaut of Daimler-Benz.

In each case, the great man takes all the credit, and also all the responsibility. In rare cases, however, we find a great man who bears the responsibility and drains his own energies, while freely acknowledging the brilliance of the men who work under him; and in very rare cases, it is recognized by them and others that he leads that team by virtue of being the most brilliant of the lot.

Most rare of all is the solitary case of such a man bringing about a worldwide revision of the principles and techniques involved in car design and manufacture, of leading the industry and the people out of one era and into another. That man was Carlo Cavalli, and the team he led was the staff of what was probably the most gifted such establishment ever known in the whole history of the car industry, the Technical Office at Fiat in the years 1919-1928.

Cavalli had been at Fiat since 1905, and it was a true case of a man finding his real vocation. Scion of a long line of judges and other lawyers, he had been brought up with the intention of maintaining that tradition; and because his father wished it he did graduate in Laws. His real love was engineering, however, and when in his late twenties he joined Fiat in their Technical Office, he immediately gave proof of his exceptional abilities.

He was a magnificent draughtsman, which in those days and in that milieu carried (as we have noted) a good deal more weight than an engineering degree. Fiat's chief engineers (apart from the first, Aristide Faccioli, who did not last long) belonged to a class established by Faccioli's 1901 successor Giovanni Enrico, who had been director of the Rome power station. Narrow specialists were deemed inadequate: A real engineer should be able to study the problems of any kind of design work, and to solve them sensibly.

Cavalli was just such an engineer; he was also a gentleman with many other admirable qualities, and he soon made a very good impression on the man above him in the hierarchy, Guido Fornaca. Some historians describe Fornaca as the chief engineer, but in fact he was head of the Technical Office in that he directed its policies in the commercial rather than the engineering sense. Fornaca saw in Cavalli a very reliable assistant: Here was no jumped-up mechanic off the factory floor, but a thoroughly professional gentleman, honest, scrupulous, displaying clarity of thought and dignity of bearing. Combining the shrewd and perceptive brain of a lawyer with the talent and enthusiasm of a creative engineer, he showed himself extraordinarily versatile.

It was an era when engineering was running riot in all manner of new directions, and Cavalli helped Fiat on their way to significance in most of them. He played a big part in the design of the heavyweight racers that were so successful during the next few years (especially in 1907), and of their first shaft-drive cars—and also saw to the first aero-engines, to airship engines, marine motors, military vehicles, agricultural tractors, and other ventures which Fiat were considering. In 1919 he was appointed technical director of the Technical Office: He was 41 years old, a new formula was about to govern the rules of motor racing, and the time was adjudged ripe for Fiat to return to that field and make some convincing points.

Cavalli did just that, masterminding a sequence of utterly brilliant and extremely advanced Grand Prix racers which effected a complete

Page 333 Dante Giacosa. Note the fine-drawn features, the carriage, and especially the hands. (Photo courtesy of Fiat)

Page 334 Avv Carlo Cavalli (*top left*) set the Fiat design standards which Giacosa inherited; Ing Cesare Cappa (*top right*), considered 'fanciful' there, tended to flout them. Ing Ettore Cordiano (*bottom left*) is an all-around engineer particularly creative in suspension design, whilst engines specialist Aurelio Lampredi (*bottom right*) may be the greatest designer not to have a chapter to himself in this book. (Photo courtesy of Fiat)

revolution in design. They did not enter the fray very often, but whenever they ran they left any man of understanding so convinced of their conceptual superiority that there was nothing to do but to copy them; and that, for decades and generations to come, is what practically everybody did.

Some of those Fiat racers simply devastated the opposition. Others made what were little more than demonstration runs. Sometimes the less astute onlookers missed the point that was being demonstrated, but sooner or later and in one way or another everything that Fiat did was copied.

This is no place for a detailed account of the racers' designs and fortunes. Suffice it that Cavalli showed the world the high-speed engine with twin overhead camshafts operating inclined valves in hemispherical combustion chambers, with profuse and precise support of a straight eight- or six-cylinder crankshaft and all other moving parts in properly lubricated rolling-element bearings. He shook the world with the first supercharged GP car. He beautified the world with the first car to be designed (in 1922!) with the aid of a wind tunnel so as not only to minimise drag but also to eliminate aerodynamic lift. As for the more immediate needs of the world, what many people would have considered the real world, Cavalli provided for them too: He designed the 18NL and 15Ter trucks, the prewar Taunus and Zero cars, and the 501, the first to be built in really large numbers after the Great War. While his racers sparkled with iconoclastic innovations born of scientific logic, his production cars bloomed as economic expressions of practicality bred of the human condition.

He did not achieve all this by himself. As we have seen, he had under him a superb team—superb because he himself was an inspiring leader, mentor, guide, monitor, orchestrator and conductor. Flourishing in that intellectual hot-house were the brains and drawing hands of Bazzi, Becchia, Bertarione, Cappa, Massimino and Zerbi as design engineers, while in charge of car preparation and team administration was Jano. Nobody could compete with such a team; and so the headhunters moved in, emissaries of shifty French and jealous Italian rivals. Within a few years all but Cavalli and Zerbi deserted Fiat.

In doing so they carried the Fiat creed, largely the creation of Cavalli, into the design offices of most of the famous and influential manufacturers of Europe, whence their work spread out all around the world to colour the imaginations of even the isolationist Americans. The family tree they thus created is intricately branched, but as each individual moved from one employer to the next, sometimes picking up an acolyte on the way to whom he imparted the original tradition, the ramifications grew amazingly comprehensive.

Bertarione was the first to go, creating the Sunbeam (nicknamed the 'Fiat in green paint') which won the 1923 GP at Tours. The next seducer was Alfa Romeo, and as soon as Luigi Bazzi was safe there (since he was a friend of Enzo Ferrari) he got them to woo Vittorio Jano away. The things Jano did for Alfa Romeo were even more memorable than those he later did for Lancia; and meanwhile his other Fiat inmates were filtering off to work as the years went by for Talbot, Itala, Lorraine, Ansaldo, OM, Darracq, Breda, Piaggio, Hotchkiss, Lago, Maserati, Ferrari and the Lord knows whom else. The firms who did not seek to employ them were those who felt that they had the ability to copy them, as was usually the case in Germany and Austria, more occasionally in England and America.

Tranquillo Zerbi remained, switching between astonishing GP car engines and record-breaking aircraft engines. He it was who took over when, in 1928, Cavalli had to retire. The latter had worked himself as hard as only a man of principle can; now, illness and exhaustion forced his retirement to the valley of Vigezzo where he had been born in 1878. Nineteen years later he died, at Santa Maria Maggiore in Novara, and by that time the roads of his native land and many others were running with evidence of how well he had established an engineering creed.

It had been well followed: In the very year when Cavalli retired, a newcomer joined Fiat to begin a career as distinguished as any. That newcomer was born in the very year when Cavalli joined Fiat, and his first full year as director of the engineering division was the year in which Cavalli died. When one door closes, another opens; the young man who walked in when Cavalli bowed out was Dante Giacosa.

Giacosa's fresh and inventive spirit, after brief observation in a department dealing with military vehicles and other load-carriers, was found a niche in the aero-engines division, where under the direction of the gifted Tranquillo Zerbi (who had a year earlier finished the last and most brilliant of Fiat's Grand Prix racers) some remarkable advances were being made. It was no less remarkable, as an instance of the company's ability to give the right man the right job, that Giacosa was still in the aero-engines department when, at the age of 28, he was entrusted with designing the mechanical content of the everlastingly famous Fiat 500.

The engine of this little marvel was to emerge no larger than the carburettor of the aero-engine upon which Giacosa had been working; but the engine was neither the beginning nor the essence of the 'Topolino,' as the car came to be nicknamed when it was launched into a world already enjoying Mickey Mouse. It came on the scene as the first microcar to be properly refined and the first to be free from savage compromises of habitability or roadworthiness. What made it the more remarkable was the vital stipulation in Giacosa's brief from Senator Agnelli, passed on by the then director of the aero engines technical office, Antonio Fessia: It was to sell at 5000 lire. Italy was rejoicing in the availability of the 1932 Balilla, available at 11,250 lire for the two-door version or 12,950 for the four-door—and here was this young fellow, recently promoted to departmental head at a monthly salary of 1500 lire, and chosen to open new doors to Italian mobility.

In those days car bodywork was designed by a different department, a relic of the old coachbuilders' era, and Giacosa had to accept without much change (and, to be fair, without much quibble, since on the whole he liked it) the design prepared by the resident coachwork engineer Rodolfo Schaeffer. Clean and curvilinear, it echoed strongly the modern lines of the Fiat 1500 saloon which was to set new standards of stream-lining for production saloons when it appeared in 1935. The Fiat 500, however, was to be in some ways even more clever: Giacosa determined to design its chassis in such a way that it would only develop its role as a structurally stiff base when attached at numerous points to the metal bodywork of Schaeffer's design. It was not yet time for Fiat to enter the ranks (sparse as yet) of makers of 'monocoque' hulls, but he was anxious to take all possible steps towards that goal.

His aero-engines background encouraged him to hang the engine in the extreme front of the sloping-nosed body, cantilevering it from the structure supporting the independent front suspension. Dimensioning the engine to fit into the tiny space available gave him sleepless nights, but after trying episodes with noisy bearings and low-octane fuel he arrived at the 569cm³ side-valve four-cylinder water-cooled job whose 13bhp was not out of keeping with a wheelbase no longer than two good paces.

Work was intense, furiously concentrated, with deadlines looming; and all the time Giacosa's insistence on keeping everything as simple as possible goaded him to try once more to eliminate anything that was either deleterious or superfluous. In the end he was proud and satisfied: The new model marked the start of true mass-production at Fiat, the very reason for the construction of the wonderful Lingotto factory with its banked test track on the roof of the five-story building, half a kilo-metre long. Soon a hundred Topolini were emerging daily, each with room for two people and 110 lb of luggage, and capable on its newly commissioned 15-inch Pirelli tires of at least 53 mph. It was heavier than had been hoped, at 1190 lb, but that was due to it being a more complete and practical car than originally envisaged. It was also more expensive, at 8900 lire; but that was inflation.

With the success of the 500 clear, Fiat streamlined the management system of the Technical Offices, giving Fessia more power and putting Giacosa immediately beneath him as engineering manager (coachwork was still separate) of the cars department. Yet, even more important than the technical direction given by Zerbi (and by the complex Fessia, the intermediary between him and Giacosa) was the overall management progressively exercised by a man who had joined Fiat in 1921 and was managing director from 1928 to 1967. Professor Vittorio Valetta, a diminutive but authoritative business graduate, was the strong man who inspired and guided the development of Fiat, rebuilt it after the World War, and perpetuated the creed of its creators by making it a policy not to confront the people with the future but to lead them gently into it. The technical advances evident in the cars created by Giacosa were always consonant with that overriding principle.

Sometimes Giacosa chafed at the restrictions imposed by the conservative element in the firm's management. In particular he suffered for a long time from the iron inflexibility and paralytic caution of the production manager, Commendatore Alessandro Genero, for the production engineering department frequently exercised its prerogative of overruling the men engaged on design and development. Time and again some Giacosa felicity was thrown out in this way, yet Giacosa continued to admire Genero as a man and accepted the authority invested in him.

It was for years to come a matter of biting the bullet, of saving up design principles that he recognised as good until such time as they could be imposed and accepted with a good grace. Perhaps the most trying episode was during the design of the Fiat 1400, the first 'postwar modern' to be undertaken by Fiat and the first car for which Giacosa, now director of the engineering division, was wholly responsible.

The management seemed to expect that the 1400 should be all things to all men, and to be blind to the impossibility of realising that ambition. It had to have an engine no larger than 1300cm^3, to avoid a tax step which would reduce its appeal to the domestic market; it had to have such styling and size and furnishings as might make it appeal to the Americans, whose money was being borrowed to put Fiat back on the map after Allied bombers had made such an effort to remove it during the war. It had to be no bigger overall than the Fiat 508C, as big inside as a De Soto, and to offer such performance as might endear it to users worldwide.

In vain Giacosa argued that what was wanted would be better achieved in other ways. The greatest concession he could wring from blue-blooded Luigi Gajal de La Chenaye, chief of the sales department for as long as Giacosa had been at Fiat, was to increase the engine to 1400cm³. Genero and others put obstacles in his way at almost every step; Giacosa loyally and patiently treated each of these as a challenge to his professional skills, and did the best he could.

The result was Fiat's first car to have a stressed-skin hull with no separate chassis, the first to have an engine with stroke markedly less than the bore, the first and only to have a hydraulic coupling between flywheel and clutch (to solve a vibration problem for which an effective crankshaft damper could not be developed in time); and when, shortly after the 1950 launch of the 1400, Giacosa came up with a longer-stroke engine which turned it into a 1900 of perfectly acceptable performance, it was to be a great success. The fact that many of its mechanical components were adopted for service in early Ferrari models may or may not be a compliment to Giacosa's work; it depends on what one thinks of early Ferrari models . . .

The 1400 was only begun after a lengthy investigation of the front-wheel-drive Aluminium Française Grégoire prototype in which Simca were taking an interest that might usefully extend to Turin. Giacosa admired the Grégoire part of it, but took a dim view of the Aluminium part: He was able to demonstrate (as quite a few aero-engineers were having to do in assorted dens of misplaced enthusiasm around the world at that time) that an equivalent weight of steel correctly used would do the job better, longer, and more cheaply, than all that light alloy. On the other hand, he would have loved to make a start on front-wheel drive, if only decent drive-joints had been available.

Giacosa would have liked the Topolino to have had front-wheel drive, but (not least because founder Senator Agnelli had been a passenger when a 1931 front-drive prototype by Oreste Lardone suffered an accident) he had to argue for thirty years before winning his case. When the Senator (who very nearly caught fire in the accident and who sacked Lardone on the spot) was dead, Giacosa contrived to have engineer Lardone, whom he considered very able and unjustly treated, brought back to Fiat. Then, first with the Autobianchi Primula and then with the Fiat 128, he established the definitive form of front-wheel drive, with transverse engine and separate but parallel transmission, and with suspension designed to counter the naturally deleterious effects of front-wheel drive upon the handling.

Despite the efforts of innumerable protagonists, ranging from Christie in the USA to DKW in Germany, and not forgetting French contributions ranging from Cugnot in the 1760s to Grégoire in the 1920s, it may be said of front-wheel drive that Citroën made it interesting and

Issigonis made it popular. Giacosa made it universal.

There is not the slightest exaggeration in the assertion that it was he, with the brilliant 128, who established the definitive form that has become hallowed by emulation everywhere. It was not just a matter of setting the engine transversely across the nose, although this was an early proposition of his. It was not a matter of making room for such a proposition by devising a strut form of independent front suspension, though this too was something that he thought of long before Mac-Pherson. It was also a matter of arranging a three-shaft transmission layout which kept the gearbox out of the sump, allowed the engine to be set lower, allowed reasonable access to the clutch and ancillaries, and enabled both drive shafts to be endowed with matching torsional stiff-ness (despite their asymmetry) as a first step towards eliminating spurious steering effects.

Then there was the use of inboard pot joints, of transverse rear springs allowed to flex by their supports in such a manner as to be stiffer in roll than in bump (and thus eliminate the cumbersome anti-roll torsion bar and its linkages), and particularly the use of strut suspension at the rear, steered by fixed links which altered wheel alignment to compensate for the effects of suspension deflexion induced either by bumps or by cor-nering forces. Giacosa credited this rear suspension to Ettore Cordiano, an amiable engineer of wide-ranging abilities who played a significant role in the development of some of Fiat's most enchanting cars in the late 1960s and early 1970s; but it is clear that Giacosa, recognising how important the 128 would be, was masterminding everything.

The same was true in the engines department, where the renowned Aurelio Lampredi, erstwhile saviour of Ferrari, evolved a little paragon of an engine that was not only to prove capable of amazing development but also, in its original sweet smooth and softly singing guise as a mass-production touring unit, made it finally and undoubtedly absurd for the camshaft of any engine to be anywhere but overhead.

Giacosa encouraged that, too, and much else. Always a stickler for lightness, he encouraged the chassis designers to prune the unsprung mass to a mere 55 lb at each rear corner. Always concerned to eliminate the vibrations which derive from anything running 'out of true,' he had the wheels spigot-located on their hubs, the nuts serving only to stop them falling off again. Such admirable details abounded in the 128, making it a car which brought to its category a refinement that had previously been unknown.

Twenty years and more later, these felicities are commonplace, and we may have forgotten how rare and special the 128 was when it was new in 1969. We may recognise that some details have been improved

Page 341 This 1932 Balilla (*above*) was Fiat's 'car for the people' when Giacosa was given his chance to design the Fiat 500. His next task was the 508C (*below*) with overhead valves and an all-steel body, sire of a long line of famous and influential Millecenti. (Photo courtesy of Fiat)

Page 342 The Fiat 1400 (*above*) of 1950 was an exercise in compromise. The Cisitalia (Pininfarina's 1949 coupe version is shown below) was a triumph of purification, based on Fiat 1100 components. (Photo courtesy of Fiat)

Page 343 Scarcely longer than Giacosa's original 500, his 600 (*above*) of 1965 was a true four-seater in miniature. His 1957 New 500 (*below*), with two cylinder rear engine, was a minimalist expression of similar principles. (Photo courtesy of Fiat)

Page 344 In the Autobianchi Primula (*above*) of 1964, Giacosa's layout for front-wheel-drive engine and transmission was given an airing, pending the arrival in 1969 of the epoch-making Fiat 128 (*below*) which set the pattern for all popular cars in subsequent years. (Photo courtesy of Fiat)

since. We may even find an occasional popular (in the European sense implying small) car which does not clearly derive from it; but we should have to look jolly hard to find any mass-production front-drive car that does not in a multitude of ways echo the themes that Giacosa composed.

Devoted though he was to the real needs of the common motorist, Giacosa enjoyed letting his hair down: A hair-raising high-speed thrash in some prototype over the most testing roads in the Alpine foothills was a favourite way of letting off steam whenever the production engineers seemed particularly oppressive—as, to be fair, it was their job to be. So when, late in 1944, he was invited to design the Cisitalia racer as a spare-time consultant for the enthusiastic Piero Dusio, he knew just what properties it should have to succeed after the war as a limited-budget competition car.

He also entertained some notions of other features that might prove desirable, such as a gearshift which needed no skill and constituted no distraction. He devised one which was operated automatically by use of the clutch pedal, a small selector lever governing preselection of bottom gear when necessary. Unfortunately his apparatus could only work with a three-speed gearbox, and for this little racer that was not enough. Nor would it be when, later, he toyed with adopting it in the Fiat 1400. On the other hand, the rear suspension, locating the live axle by a torque linkage and reversed quarter-elliptic springs but suspending it by helical coil springs, was with relatively little modification to prove quite appropriate to the production saloon.

Giacosa had been bombed out of his home during the air raids of 1942, and had gone from pillar to post in search of accommodation ever since. Now, working every night and throughout each weekend in the lovely villa that Dusio had put at his disposal, he was able to seek intellectual refreshment after the factory slog finished at 5 o'clock each working day, sitting at a drawing board which was subject neither to interruption nor to overruling. Giacosa followed his instincts and sought to achieve as much as possible with as little as possible. Minimal cost, minimal weight, minimal size, minimal waste, were all part of the Giacosa ethos—but never a false economy. Safety, durability, and beauty (engineering and art, he insisted, were inextricably linked), all were essential to this most humanitarian of automobile engineers.

The front suspension was adapted from the Fiat 500; the 1100 engine was tuned from its normal 32 bhp to yield 62. With its then ultra-modern multitubular chassis (built, appropriately, in what had been a bicycle factory) and its shrewd use of Fiat production parts, the light-weight Cisitalia 1100 proved how well Giacosa was matched to the task; and when his two-seater version finished second, third and fourth in the

1947 Mille Miglia, with Nuvolari's car actually faster (after a 27-minute electrical delay it finished second by just 17 minutes) than Biondetti's winning 3-litre Alfa Romeo, Giacosa might have been thought misplaced at Fiat. But he had done some aerodynamic work for Fiat on a sporting two-seater version of the Fiat 1100, and that finished in 5th to 8th places, ahead of all the other more powerful cars—evidence of how good his production machines were as 'driver's cars.'

That Fiat 1100 was, second only to the 128, possibly his most influential creation. He did it immediately after the 500, accepting (as in the case of the 500) stylistic guidelines that had originated with the refined six-cylinder 1500 of 1935. Mechanically, however, he conceived it as a further big step ahead from the 1932 Balilla 508, itself a car of remarkable vivacity and technical refinement in its time and class. Giacosa's 508C of 1937, with independent front suspension (he wanted it at the rear too, but had to wait) and hydraulic dampers, with good hydraulic brakes and a big-bore short-stroke engine of 1089cm^3, attracted the name Millecento (1100); and that name stayed in the Fiat catalogues for thirty years.

By the standards of its time, that car was a prodigy in handling, road holding, ride, performance and comfort; it introduced people to the experience of long fast economical journeys on the spreading Autostrade (which Italy had begun to build in 1923, a decade before Germany), and to the notion of a fast-revving engine resisting wear better than the old long-stroke strugglers. It confirmed its principles as a car built for the French under the Simca banner, so that Gordini in turn could build effective cut-price racers from it. Most of all, it showed a post-war generation (newly interested in motor vehicles by the most mechanised war yet seen) that the way to go was by pursuing roadworthiness, economy, and speed, with equal zeal for overall social efficiency.

That generation was no less impressed by Giacosa's little rear-engined jewel, the four-cylinder 600 which begat the 850, a swarm of high-performance Abarth derivatives, and the Nuova 500. Even before Giacosa had been made chief engineer in 1946, he had many long-range projects in mind. He knew that rear-engined cars would not long prosper; he monitored the creation (largely by his assistant Oscar Montabone) of the conventional 124, a car which was almost outrageously competent and outstandingly safe, and was to endear itself to Russians and racers alike—and at the same time he masterminded the creation of the front-drive design which was to dominate generations to come.

There were also many other designs, such as the sporting 2-litre 8V, the homologation-special Dino 2-litre with which Fiat sought to give Ferrari a helping hand, the magnificent Dino 2.4, the patrician 130, and more all passing beneath his sensitive scrutiny. Sometimes the sensitivity amounted almost to pain: Giacosa admitted that he could never actually enthuse about the 130, for he believed to his very soul that a car should be small and as simple and affordable as it could be. He also recognised some truth in the old Fiat notion that, for the rough roads and demanding drivers of Italy, a car had to be made more robust than might be acceptable elsewhere. This was why big Fiats had in the past been underpowered and overweight compared with the American models seen as exemplars of practical big-car philosophy; nevertheless the 130 finally emerged as a superbly detailed and impeccably behaved luxury car. Some people sneered at it as 'a Ford made by Mercedes-Benz,' not realising what a compliment they paid it; a Mercedes-Benz made by Ford would have given them something at which to sneer. If it was not a success, it was because too few people understood the 130; the loyal and scrupulous Giacosa understood, even if he did not sympathise.

By this time he was no factotum: He had some superb assistants, Lampredi and Cordiano coming quickly to mind. He was open-minded enough to allow for all ideas: During the war, he had spent much time musing about the possibilities for new cars opened up by the technological developments that the war itself accelerated. Afterwards he applied himself to all possible new ideas that might come within his ambit: gas turbines, hovercraft, alternative fuels, automatic transmissions, all came under his scrutiny—which meant study, design, and experimental manufacture. It is remarkable how practical all this research was, as though Giacosa had taken to heart the old French adage 'Inventir n'est rien; construire, c'est peu; essayer, c'est tout!'

After years being surrounded by destruction, such positive activity must have been heartwarming for a man whose very nature was, however carefully cool of head, to be warm at heart. He had spent the war years hounded by bombings, doing perfunctory adaptations of vehicles for the army, designing an aero-engine that he felt sure would not be needed (he was right), refining suspension designs (hampered by the occasional destruction by bombing of his experimental cars), and generally being reminded of the straitened and strictly economical circumstances in which he grew up and was moulded for his future. In the summer evenings he and electrical wizard Marchisio (later a founder of FISITA) took a low-geared ATV up into the hills to find rare mountain flowers: Marchisio wanted to establish a collection and have Giacosa draw them. You see (and you can see, in much of his work), with all his other gifts and sensibilities, Giacosa was also an artist.

12 COLIN CHAPMAN

by PHILIP TURNER

COLIN CHAPMAN (1928–1982) English: An outstanding designer, driver, industrialist who from an Austin Seven Special constructed in his girlfriend's garage created a Grand Prix team that won six World Championship of Drivers and seven World Championship Constructors' Cups with the cars of his design and built up one of Britain's leading specialist car producers, again with outstanding road cars of his design.

COLIN CHAPMAN

ANTHONY COLIN BRUCE CHAPMAN was a fascinatingly complex character, a man with a dozen different and conflicting reputations. That of
being a hard man, with perhaps a soft centre; of being the relentless
pursuer of the fast buck and yet a man of his word. Yet on one point
alone there will be no disagreement: that Colin Chapman was a truly
great designer of racing cars and of high-speed road cars.

He was born in Richmond on the west side of London in May 1928,
but two years later his father, Stanley Kennedy Chapman, moved the
family to North London when he took over the Railway Hotel at
Hornsey. It would have been only natural if young Colin had become
a keen train-spotter, for the 'Flying Scotsman' in all its glory of streamlined steam power used to thunder past the hotel every day as it picked
up speed after leaving King's Cross en route for Edinburgh. But instead,
Colin Chapman was far more interested in the cars that swooped round
the rather tricky bend just outside the hotel's front door.

However, the main railway line to the north was a prime Luftwaffe
target during the war years, but fortunately they missed the Railway
Hotel, from which, in fact, the family moved just before the end of the
war to a house in Beech Drive, North Finchley. Mr Chapman senior,
however, retained his ownership of the Railway Hotel, a fact which
played no mean part in the early days of Lotus. In 1945, Colin Chapman
entered University College, London University, to study engineering,
commuting at first between home and college on a 350cc Panther
motorcycle. In November of that year he smote a taxi and demolished
the motorcycle, but fortunately not himself. No doubt, however, his
parents regarded the strips of plaster adorning their son as so much
writing on the wall and decided that he might live longer if restrained
to four wheels, so that Christmas they presented him with a 1937 Morris
Eight Tourer. Oddly enough, about this time I was motoring in a 1938
version of this model—it was even the same colour, maroon—so I can
appreciate the part it played in developing Chapman's driving skill.

It was while he was still a student that Chapman, in association with
a fellow student, Colin Dare, began dealing in used cars of the very
second-hand variety. Chapman had sufficient self-assurance to deal in
the Warren Street jungle of those days which was inhabited by all the
sharpest of traders, a very remarkable state of affairs for a young student,
especially as he made money out of his deals there. The honey-tongued
persuasiveness that Colin developed through this wheeling and dealing
remained with him for the rest of his life.

But alas, when the basic petrol ration was abruptly cancelled in October 1947, owing to yet another financial crisis, used-car prices nose-dived, and the Chapman-Dare partnership found itself holding a stock of cars for which grossly inflated prices had been paid. This brought about the abrupt ending of Chapman's interest in used-car dealing and the stock was unloaded on to the depressed market at a considerable loss. If the Government had not been ruining the motor industry as usual, one is tempted to wonder whether Chapman, the used-car dealer, might not have gone from strength to strength to become one of the pillars of the motor trade, with a chain of showrooms. Instead, Chapman left the car trade and went in for motor sport, spending long hours converting a 1930 Austin Seven, last survivor of the second-hand car stock, into a trials special. This conversion process, which began with the removal of the car's fabric saloon body, took place in the lock-up behind the house at Muswell Hill of his girlfriend, Hazel Williams.

The body that replaced it was a very clear indication of things to come, for in its construction Chapman applied aircraft principles in order to obtain a structure that gave additional stiffness to a chassis frame that was all too flexible, in spite of the channel-section side members having been boxed-in. Chapman had joined the University of London Air Squadron, thereby giving vent to his interest in aircraft which always ran side by side with his passion for cars, and it was no doubt this close association with flying that set him thinking along aircraft construction lines. The body for the trials car was made from alloy-bonded plywood on a stressed framework with three bulkheads.

No major modifications to the suspension or engine were made at this stage, though later the beam-type front axle was converted to independent front suspension by cutting it in half and pivoting it in the middle. This form of ifs was to feature on many future Lotus cars—and to inspire the ifs for the Hillman Imp, for which Lotus-owner Mike Parkes was so largely responsible. When completed, the rebuilt Austin was re-registered, not as an Austin special but as a Lotus, the very first of that distinguished line. The origin of this name was to remain a private—and probably rude—joke of Colin Chapman and Hazel. The first Lotus was run in comparatively few trials in 1948, its first season, for Chapman was taking his final examinations at London University from which he emerged triumphantly as Colin Chapman BSc (Eng).

The year 1948 saw not only the completion of the first Lotus but also its creator joining two organisations which were destined to have a considerable influence on his future. One was the Royal Air Force which he joined on a short-term flying engagement to do his national service. The other was the 750 Motor Club. Chapman had already reached

Elementary Flying Training standard in the Tiger Moths of the University of London Air Squadron and had thirty-five hours of solo flying in his logbook, so he was sent to Tern Hill, where he graduated to Harvards.

The 750 Motor Club is almost unknown to the general public, yet this body has certainly exerted more influence on racing-car design than the Institution of Mechanical Engineers. The club was founded to cater for enthusiastic owners of Austin Sevens. Now, for a would-be automobile designer who wishes to experiment, the Austin Seven had many advantages. First, it was fairly cheap. Second, there were numerous bits and pieces available for modifying the basic concept of the car, and third, the roadholding in standard form was so diabolical that any modification was almost certain to bring about an improvement—very encouraging to a new designer. In the 750 Motor Club, Chapman made contact with numerous like-minded enthusiasts with whom he could discuss problems of design.

It was during his RAF period that Chapman designed and built the Lotus Mark 2. Although it was again based on the Austin Seven chassis frame, it was the first Lotus to be Ford-powered, initially by a Ford Eight engine which Colin managed to exchange for a Ford Ten unit by one of those fantastically complicated Chapman-type deals which, in true Chapman fashion, concluded with him owning a better engine plus £5 profit. The new Lotus was designed to compete in circuit races as well as in trials, and although it proved its worth as a trials machine by winning some events outright in the opening months of the 1950 season, the car really attracted attention in a big way when Chapman went racing with it.

By this time Chapman was a civilian once more, for he left the RAF after gaining his wings, having decided against taking a five-year permanent commission. Perhaps this was just as well, for otherwise he would probably have been shot at dawn for telling Air Marshals how to run the RAF.

In April, Chapman joined the British Aluminium Company, as a construction engineer, principally engaged in suggesting how things could be made much better and lighter in aluminium than in old-fashioned metals. Was this the start of Chapman's quest for lightness and yet more lightness which, in later years, was to horrify the British components industry, a quest which he continued to pursue with unrelenting vigour?

In spite of the new job, Chapman was too short of money to go to the British Grand Prix at Silverstone in May, but in June he drove a Lotus to its first race victory when the Mark 2 won the scratch race at

the Eight Clubs' Silverstone meeting after a tremendous battle with
Dudley Geoghagan's Type 37 Bugatti. Already, Colin Chapman the
racing driver was showing just as much promise as the Lotus car.

At the end of the 1950 season, Chapman decided to concentrate on
racing rather than on trials, and therefore sold the Mark 2 to Michael
Lawson, who scored many successes with it in trials. To replace it,
Chapman built the Lotus Mark 3, which was specially designed to
compete under the new 750 Formula which the 750 Motor Club were
promoting. Once again, an Austin Seven chassis—plucked from a 1930
saloon—formed the basis, stiffened by the usual boxing-in of the side
members and also by a 15 gauge tubular steel hoop at the scuttle and
also by a triangulated tubular structure above the engine, with the base
of the triangle attached to the scuttle and the point to the top of a tubular
steel hoop at the front of the car, just behind the radiator. Thus did
Chapman take his first tentative step towards a space frame. A Ford
Eight beam-type front axle pivoted at the centre, long Newton telescopic dampers and Lockheed hydraulic brakes in place of the normal
Austin 'press and hope' mechanical type were other chassis modifications.

Some automotive designers are chassis men pure and simple with
scant interest in engine design, while others are engine men, only too
happy to leave the running-gear to somebody else. But the slender
resources with which Chapman was working forced him to become just
as enterprising over engine design as over the design of the car as a
whole. In fact, the Mark 3 Lotus probably owed as much if not more of
its success to its excess of power over its rivals as to its outstanding
roadholding.

With considerable ingenuity, Chapman converted the normal two-
port Austin cylinder block, with its siamesed inlet ports, into a four-port
block by building a special inlet manifold from welded sheet steel. Each
branch was divided into two by a vertical steel strip which projected
from the end of each manifold into the inlet port cast into the block. A
massive twin-choke Stromberg carburettor from a Ford V-8 was mounted
vertically on the manifold. An exceptionally slim light-alloy body
weighing only 65 lb meant that as little power as possible was required
to push the Mark 3 through the atmosphere. With this car, Chapman
had a tremendous season in 1951 and the little car was quite unbeatable
in 750 Formula races. It even proved a tough opponent when matched
against cars of much greater capacity.

It was, in fact, the success of the Mark 3 which turned Lotus from a
hobby into a business, for fellow 750 MC members were asking Chapman to build them replicas, or at least to supply them with components.
Michael Lawson, too, asked for a replacement for Lotus Mark 2 to cope

with changed conditions in the world of trials.

Chapman had built the Mark 3 in co-operation with two keen brothers, Michael and Nigel Allen; in fact, the car had been built in their private garage. Michael Allen agreed to join Colin Chapman in turning Lotus into a business so, on 1 January 1952, the Lotus Engineering Company came into being, with Michael Allen working full time and Chapman in his spare time from his work with the British Aluminium Company. It was indeed fortunate that Chapman senior still owned the Railway Hotel in Hornsey, for at the back of the hotel was an old stable, a left-over from the time when every inn of any size had to provide stabling for the horses of its patrons. The old stable therefore became the 'factory' of the new Lotus company.

The new company began life by building a Mark 3B 750 racer for a fellow 750 MC member and an Austin Seven-based Mark 4 trials car for Michael Lawson. Mark 5 should have been a road-going sports-car version of the Mark 3 but never got beyond the design stage. Instead, Chapman embarked on the design of the first production Lotus, and the first not to be based on an Austin Seven chassis.

It was the Lotus Mark 6 that truly founded the fortunes of Lotus as a company. Owing to lack of space in the Hornsey stables, there was no question of Lotus setting up a production line to build cars in the normal way. What Chapman therefore did was to design a multi-tube space frame to which modified Ford parts could be added by the owner, who thus bought his car in the form of a kit to put together in his own garage. As a result of Chapman's detailed and correct stressing of the structure, the space frame was very rigid and exceptionally light, weighing only 55lb. The basic space frame was rendered even more rigid by riveting to it the stressed sheet aluminium panels that formed the floor, scuttle and body side panels. Among the Ford parts was a beam front axle cut in the middle to provide swing-axle ifs. This arrangement kept the front wheels nearly upright when cornering at speed, thanks to a low roll centre, and enabled relatively soft suspension with considerable travel to be obtained, a feature of most subsequent racing Lotus models.

The Mark 6 made its first appearance at the MG Car Club's Silverstone meeting in July 1952, where it went well and secured a couple of second places. In September, it became the first Lotus to be entered for an International meeting, its entry being accepted for the *Daily Mail* 100-mile sports-car race at Boreham—now the home of the Ford competition department. But an off-the-course excursion during practising necessitated a return to the Hornsey workshop for repairs and, while being driven there by Nigel Allen, the prototype was completely destroyed by a van which shot out of a side turning and clobbered it.

Nigel Allen was unhurt, and the van driver was dealt with in court but, all the same, it was an absolute disaster for the young company with bills to pay, no money in the bank and a wrecked car as their main asset.

Fortunately, the Mark 6 had been comprehensively insured, and the insurance money just about paid the bills, but by the end of 1952 Colin Chapman was back to square one again. Michael Allen decided to withdraw from the company and left, taking the crashed Mark 6 as his share of the assets.

In J.M. Barrie's play, *What Every Woman Knows,* the theory is propounded that what every woman knows is that behind every successful man there is an equally outstanding woman. Colin Chapman's girlfriend from his London University days, Hazel Williams, had shared all his subsequent ups and downs. She had assisted him in the building of the Marks 1 and 2 in the lock-up behind her home, she had driven these and subsequent models with considerable spirit in trials, sprints and races, and she had acted as secretary, tea-maker and general aide to the new little company in its Hornsey stable. Now, at this crisis in Chapman's life, she lent him £25, and with this sum, plus £100 from Colin, Lotus was re-formed in February 1953 as a limited company, Lotus Engineering Co Ltd.

Another attribute of so many successful men appears to be the ability to work incredible hours without collapsing under the strain. Colin Chapman would dart into the Hornsey works first thing in the morning on his way to his full-time job at the British Aluminium Co. Then, having completed his 9:30 to 5:30 daily stint, he would dash back to Hornsey and build production Mark 6s until the very small hours of the following day. In this fashion, he built the space frames and components for the first eight production Mark 6s. The body-builders moved from Edmonton to the Tottenham Lane stables at Hornsey, where they beat panels all day while Colin constituted the night shift.

Later, Chapman was joined on a part-time basis by Peter Ross and Mac Mackintosh, who both worked for the de Havilland Aircraft Co at Hatfield. They told another de Havilland man, Mike Costin, about this Chapman character, and the two met at the May meeting of the 750 Motor Club. As a result, Costin, too, came along to Hornsey to work there in his spare time. By the end of 1953 the company was on its way, for the Mark 6, in private hands and with a wide variety of engines, had achieved a great deal of success, owing to the way in which its light weight and quite outstanding roadholding enabled it to out-perform all its rivals.

But the Mark 6 was essentially a club racer for the private owner. Chapman now wished to enter the international field with a car that would be competitive in the then very popular 1½ litre sports-car class. Lotus certainly could not afford to design and develop their own racing engine, therefore performance would have to be sought in some direction other than sheer horsepower. Fortunately, Mike Costin's brother, Frank, was a most talented aerodynamicist who also worked for the de Havilland Co. His aid was therefore enlisted to design a highly aerodynamic sports car which would make up for any power disadvantage by requiring less power than its rivals to attain a similar or better performance.

The basis of the new car was a triangulated space frame of 1¼ in, 20 swg steel tubing which weighed only 35 lb and yet was remarkably rigid. Because it was built up from a multiple of small tubes, however, the task of coaxing the engine in and out of the chassis was very daunting, and in fact the cylinder head had to be removed first before the engine could be extricated.

To clothe the chassis, Frank Costin designed a most unusual-looking body which swept up from a low, sloping nose to big tail fins behind the spatted rear wheels. The aluminium body panels were attached rigidly to the space frame to increase the torsional stiffness of the car. Swing-axle ifs was retained at the front, but a De Dion-type rear axle was employed, with the De Dion tube behind the differential and located laterally by a ball-bearing in a vertical groove in the rear of the casing. An unusual feature was the use of a single coil spring in tension in conjunction with piston-type Armstrong dampers for the rear suspension. The engine was an amalgam of MG TC and Morris Ten components, with a Lucas-Laystall light-alloy cylinder head and twin SU carburettors. This unit provided 85 bhp, sufficient to propel the new Mark 8 at around 125 mph.

First appearance of the car was in the British Empire Trophy at Oulton Park on 10 April 1954, in which the car made joint fastest lap of the race with Chapman at the wheel but retired with a blown cylinder-head gasket. It had reached the start-line only after all-night work by its crew to repair frontal damage sustained when the car went straight on over a roundabout while being driven up from Hornsey.

However, Chapman won his class in the sports-car race at the Daily Express International Trophy meeting at Silverstone, and a week later it became the first Lotus works entry to compete in a Continental event, when it was driven into fourth place in the International Eifelrennen at the Nürburgring by a German driver, Erwin Bauer. The organisers would not let Chapman drive as they considered him too inexperienced, and he had insufficient time for practising to learn the Ring thoroughly. In fact, Chapman and Costin left London on Friday evening after

Chapman finished work, and so did not reach the Ring until midday on Saturday. After the race on Sunday, they caught the night boat and Chapman was back at his desk by 9:30 am on Monday morning. Later in the season, Chapman left London on Wednesday evening for the race at the Ring on Sunday before the German GP, in which the car broke its De Dion tube. This was welded and the car driven back to England in time to take part in races at Brands Hatch and the Crystal Palace the following day, August Bank Holiday Monday. Even Chapman came to the conclusion that perhaps this was a bit much.

The major success of the Mark 8 during the 1954 season was its victory in Chapman's hands in the 1½ litre class of the sports-car race at the British GP meeting at Silverstone on 17 July, when it beat the works Porsche of Hans Herrmann.

The year 1954 was notable for two other major developments in the Chapman/Lotus story. At the start of the season Chapman had decided that the racing activities of Lotus must not interfere with the money-making production side, and therefore Team Lotus was formed as a separate entity from the Lotus Engineering Co Ltd. In this first season, Team Lotus consisted of Colin as driver and a band of keen enthusiasts who gave their spare time to maintaining the Mark 8 prototype. As the season progressed, further Mark 8s were built for sale to private owners. The second major development was that, at the end of the season in October, Colin Chapman married Hazel Williams at Northaw church and the couple set up house at Monken Hadley, near Barnet, Hertfordshire. The fact that the house, like the Lotus works, had begun life as a stables, was just coincidence.

By the end of 1954, Lotus was firmly established as a going concern, with a continuing demand for Mark 6s and eager buyers seeking Mark 8s or derivatives thereof, for many people wanted to install bigger engines in a Mark 8, which was difficult without a major redesign. Colin Chapman, at the age of twenty-six, therefore decided that as from 1 January, he would give up his job with the British Aluminium Co and work full time for Lotus. Mike Costin, too, decided to leave de Havilland's and was put in charge of the development side of Lotus, becoming one of the directors of its newly formed branch, Racing Engines Ltd, a branch destined for great achievements in the years ahead.

It was just as well that Chapman could concentrate for the whole of his working day on Lotus, for he was now engaged on the design of two new racing sports cars for the 1955 season. The Mark 10 was a Mark 8 with a bigger engine-bay to house the Bristol and other 2 litre engines. The rear suspension was also modified considerably. The second new model, the Mark 9, was a smaller version of the Mark 8, for although

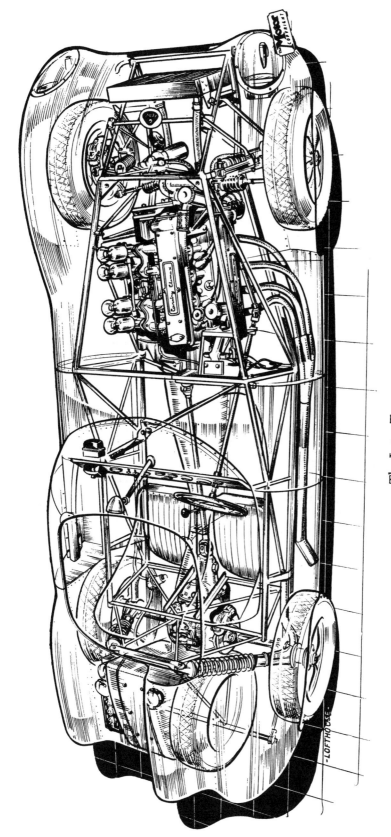

The Lotus Type 15

the wheelbase remained the same, the overall length was reduced by 2ft, principally by increasing the height but reducing the length of the tail fins.

Colin Chapman was not only chief designer, works manager and managing director of Lotus, he was also the chief racing driver of Team Lotus which, for the 1955 season, ran two Mark 9s, one powered by a 1,500cc MG engine and the other by the 1,100cc Coventry Climax engine which had made its first appearance the previous season. The original intention was that Chapman would drive the 1,500cc car and Peter Jopp the Climax-powered 1,100. At Goodwood on Whit Monday, Chapman beat Reg Parnell in a Connaught and at the sports-car race before the British GP at Aintree he won both the 2,000cc and the 1,500cc classes, and later in the same season beat Tony Brooks in a Connaught on the same circuit.

The 1,100cc car had been built primarily for Le Mans, where Chapman shared the driving with Peter Jopp. Alas, when leading the class after six hours of racing, Chapman was disqualified for reversing out of the sand bank at Arnage. However, later in the season, Chapman with the 1,100 amazed everyone in the Tourist Trophy on the Dundrod circuit in Ireland by leading all the 2 litre cars by more than nine minutes, only to be delayed for eleven minutes at his pit while a fractured oil pipe was replaced. Other Lotus landmarks in 1955 included the building of a new assembly shop at Hornsey and the first appearance of the make at the London Motor Show.

But it was the year 1956 that played a vital part in shaping the destinies both of Colin Chapman and of Lotus. It was the year which finally decided that Chapman's future was as a designer and not as a racing driver. It was the year that Lotus graduated from sports cars to single-seater racing cars. Curiously, though, the pencil-slim 1,500cc Formula 2 car, the Lotus 12, which made an entirely unexpected appearance at the 1956 Motor Show, was not the first open-wheel racing car which Chapman had designed. Way back in 1952, he had designed a racing car for the then 2 litre Formula 2 for two keen local brothers, the Clairmontes. The specification included double wishbone link independent front suspension with the coil spring and telescopic damper suspension units mounted inboard—a feature to reappear in later—much later—Lotus single-seaters, and a De Dion-type rear axle. Provisionally christened the Lotus Mark 7, the car was to have been powered by a 2 litre ERA engine, but the engine blew up in a big way before the car which was to receive it was completed. The Clairmonte brothers eventually took delivery of this non-Lotus from which the Mark 7 designation was hurriedly whipped, and the car subsequently was raced as

Page 361 (above) A youthful Colin Chapman sits proudly at the wheel of the nearly completed Lotus 1 in 1948. By the tail is Hazel Williams, later to become Mrs Chapman; (below) Lotus 2 completed in 1949 was the first Ford-engined Lotus. With Hazel navigating, Colin Chapman gained many trials successes with the car

Page 362 (*above*) From trials, Chapman progressed to racing, and by 1956 Goodwood Whit Monday meeting was capable of taking on Mike Hawthorn and beating him. Both are driving Lotus 11s; (*below*) Lotus fortunes as car manufacturers were founded on the Lotus 6

the Clairmonte Special 2 litre sports car with a Lea Francis engine under the bonnet.

Colin Chapman was also called in as a consultant to design a new space frame in 1956 for the Vanwall Formula 1 GP car, which the following year (and with Chapman-designed suspension) at last began to win Formula 1 races for Britain. Chapman was given a chance to drive in a Grand Prix—*the* Grand Prix, one might say—for at the French GP at Rheims in July 1956, he became a temporary member of the Vanwall team. Alas, very temporary, for on the approach to Thillois a front wheel locked under braking and the car headed straight for a solid concrete post at the roadside. Chapman, quite rightly reckoning that Vanwalls could be rebuilt but that Chapman was irreplacable, huddled down in the spacious Vanwall cockpit when he saw disaster was inevitable and let events take their course. Which they did—result: one bent Vanwall, one unscathed Chapman. But it is indeed fascinating to speculate on how matters would have developed if this disaster had not occurred: if Chapman had started the race, finished in the first six and continued to drive in subsequent races for the Vanwall team. For Chapman the designer has these days quite over-shadowed the fame of Chapman the driver, but in truth I believe he could well have gone right to the top in Grand Prix racing but for that locking brake at Rheims. In his day, Colin raced, and raced hard, with such drivers as Hawthorn, Salvadori, Brabham, Ron Flockhart and even Stirling Moss. And on several occasions he beat them.

The Vanwall was not the only British Formula 1 car which Chapman had a hand in, for in the following year, 1957, he was asked to revise the suspension of the 2½ litre BRM, which up to then had proved to be a somewhat twitchy motorcar. Chapman tamed it and endowed it with very reasonable roadholding. There cannot be many young men of twenty-eight who have been called in as consultant-designer by two major Grand Prix teams.

The lessons Chapman learnt from his sports cars, namely that if you have only the same engine as your rivals you must beat them by building a car with better roadholding and less wind resistance, were all applied to this first single-seater Lotus. The tubular space frame was stiffened at the front by mounting the 1,500cc Coventry Climax engine rigidly in the frame. Moreover, the engine was bolted at four points at the rear to a strong light-alloy rear engine plate which was built-in as a chassis bulkhead. Shades of the 43 and the 49 still some ten years ahead, in which the engine was to be hung on to the rear bulkhead.

The first Lotus single-seater differed in other major respects from previous Loti. Gone was the swing-axle front suspension, replaced by wishbone links, though even here Chapman novelty crept in, for the forward-mounted anti-roll bar also served as the forward arms of the upper wishbone link. A De Dion rear axle was employed when first the car was shown at Earls Court, but tests on the circuit showed the arrangement to be far from satisfactory.

To replace it, Colin developed what came to be known as 'Chapman strut' rear suspension. The strut itself consisted of a coil spring and telescopic damper suspension unit whose lower end was shrunk into a circular housing on top of an aluminium casting carrying the two wheel bearings. The struts were inclined inward, their upper ends being anchored to mountings high up on the tail behind the driver's shoulders. The rear wheels were located laterally by the drive shafts and in the fore-and-aft direction by single radius arms running forward from the light-alloy wheel bearing housing to pivot on the lower longitudinal tubes of the space frame. The Chapman strut rear suspension was not only lighter than the De Dion layout it replaced, but its geometry was so laid out that the contact point of the tyre with the road moved in a near-vertical path within the limits of usable wheel deflection.

As was to be so often the case in future, the cars, though extremely clever in design, were often let down by failures and breakages. One of the many outstanding features of the car was a Lotus-designed combined gearbox and rear-axle unit. It was an ultra compact and light five-speed constant-mesh box with an odd Z pattern gear change. The transmission was certainly ingenious but it was also fallible, the car retiring time and again with failure of the crown wheel and pinion, apparently by fault of the lubrication system, though all seemed well when the unit was tried in test rigs. A certain Keith Duckworth joined Lotus about this time and was entrusted with the job of solving the lubrication mystery. Which he duly did but, as he said somewhat loudly and insistently, even so he did not think the box would ever work. He left soon afterwards and started the Cosworth engine concern, the 'Cos' being Mike Costin and the 'worth' Keith Duckworth. Mike was persuaded to stay on at Lotus and Duckworth's Lancashire outspokenness did not stop Lotus and Cosworth from working happily together on many future projects.

The advent of single-seaters at Hornsey certainly did not mean the abandonment of the sports-car racing programme. In fact the 1957 season saw Lotus sports cars have their most triumphant run ever at Le Mans, a 750cc version of the II horrifying the French by winning the Index of Performance from the usual swarm of Panhard flat-twin-engined cars as well as its class; an 1,100cc II won its class as well and all four of the cars entered were still running at the finish.

The last few years of the 1950s were hectic indeed for both Colin

Chapman and for Lotus. The Formula 2 Lotus 12 was fitted with a 2 litre engine and run in some Formula 1 races in 1958, starting with the Monaco GP in May which thus had the honour of being the first Championship Grand Prix in which a Lotus single-seater competed— and finished, for Cliff Allison brought the car home in sixth place. In July of 1958, the first true Lotus Grand Prix car made its first appearance in the French GP at Rheims. Known as the 16, it was a slim, lovely car—and proceeded to break the hearts of everyone associated with it whether as drivers or mechanics.

The 16 looked like a miniature Vanwall, hardly surprising, as the body was by the Vanwall body-designer, Frank Costin. To keep the overall nose profile as low as possible, the four-cylinder Coventry Climax engine was tilted 60 degrees to the right, with the propeller shaft passing to the left of the driver, but with the engine mounted at this angle there was constant trouble with carburation and lubrication, so after the first few races the engine was raised until it was only 30 degrees from the vertical. Finally, the 16 was raced with the engine tilted 17½ degrees to the left. But there were other troubles, for the gearbox went through design after design and yet never functioned properly, while the car as a whole was remarkably fragile, with a horrid tendency for large cracks to form in its main structure. The cars acquired so bad a reputation for unreliability that race organizers became reluctant to accept their entries. Moreover, Colin Chapman at this time was up to his neck in other troubles, for at the 1957 Motor Show at Earls Court the first Lotus GT car, the Elite, had made its appearance.

As soon as the Lotus company began to prosper and grow, it was obvious that a proper production road car would have to be added to the range if the company was to expand at all. Even for a truly successful concern, the market for racing cars in single-seater and two-seater form was limited and the nearest thing to a road-going production Lotus, the Six, was not exactly everybody's car. In fact, it was hardly anybody's car save for the madly enthusiastic young. Even the Seven, which succeeded the Six at this period in Lotus history, was much the same only more so.

But desirable though it obviously was to add a production closed car to the range, the practical difficulties of so doing were indeed formidable. In the first place, a normal metal body was out of the question. No money existed to finance the production of the fantastically expensive dies required for the production of a pressed-steel body, and a hand-built light-alloy body would cost so much to produce that the total price of the car would put it out of reach of all except the very wealthy. Colin Chapman decided the only possible solution was to build the hull of

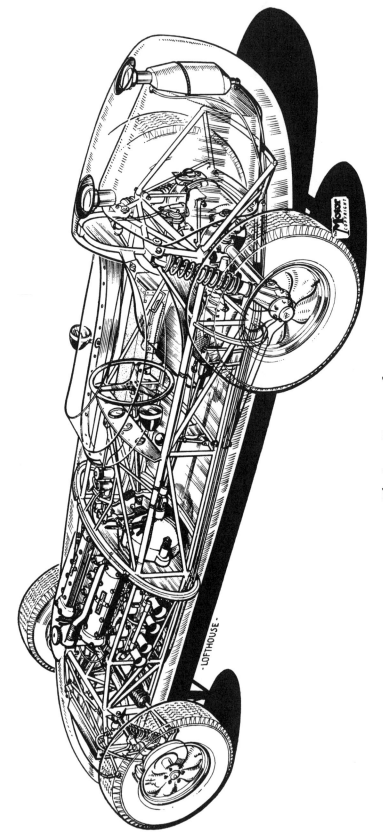

The Lotus Type 16 of 1958

the new car of glass-fibre materials, which in itself was a pretty daring step to take, for glass-fibre bodies of hideous outline, produced by knife-and-fork methods and clapped on Ford Ten chassis, had earned for all glass-fibre bodies a thoroughly bad name. Moreover, Chapman intended his glass-fibre hull to be self-supporting, with suspension and engine attachment points built into the glass-fibre shell. A friend of Chapman's, Peter Kirwan-Taylor, designed a very lovely shape for the new car; Chapman embarked on a rush course on plastic materials and their properties, and the outcome was a very advanced GT car indeed. The Elite was the first production car to use a self-supporting plastic body with plastics used as a structural material.

Tests showed the complete hull to have a torsional stiffness of 3,000lb/ft per degree, and a rather less academic test was conducted by racing driver Graham Warner when he rolled his Elite three times at Zandvoort after a rear-hub casting had collapsed in a 100mph downhill bend. When the offending hub casting had been replaced, the car appeared structurally undamaged and was driven 300 miles on the road across France before it returned to England.

The front suspension was of double wishbone type, but the front arm of the upper wishbone link was formed by the large anti-roll bar. At the rear, the Chapman strut-type suspension first seen on the Formula 2 single-seater was used, and the suspension as a whole followed Chapman's ideas on this subject by providing springing that was both soft and heavily damped. The rack and pinion steering was conventional in layout but its geometry departed completely from orthodox Ackerman principles, and in fact provided a negative Ackerman effect with the outer front wheel turning through a greater distance than the inner one on a corner. Disc brakes for all four wheels, with the discs mounted inboard at the rear, completed the specification of this little 1,300cc GT car weighing only 13½ cwt and capable of over 115mph.

The prototype Lotus Elite was completed only just in time for the 1957 London Motor Show at which it was one of the major attractions, for this good looking little two-seater coupe, with its racing-car suspension borrowed direct from the Formula 2 Lotus and its 1,216cc Coventry Climax engine, was indeed quite unlike any other car then in production in Britain.

More than a year was to elapse, however, before the Elite achieved production, for when the prototype was shown at Earls Court there was no space at Hornsey in which it could be built, nor finance available for the construction of a new plant. The promise of that prototype Elite was such, however, that finance was eventually found and Lotus acquired new works at Cheshunt, Hertfordshire, on an industrial estate to which the entire Lotus operation was transferred in 1959, no doubt to the great relief of the citizens of Hornsey who were always complaining bitterly of too much noise much too late at night.

At the new works, output of the Elite built up steadily, and more than a hundred were completed by the end of 1959, compared with only forty or so built during 1958. It was hoped that in 1960 production would build up to an output of twenty-five Elites a week, and to this end, a contract for the production of the hull was signed with Bristol Aeroplane Plastics Ltd. But alas, these hopes were doomed to disappointment, and total production of the Elite during the whole of its life was less than a thousand cars in all.

Why, then, did a design which worked so well that the roadholding not only delighted road users but also made it unbeatable in its class in sports-car racing, not pay off in substantial profits? As the wife demanded of her husband, 'If you're so clever, why aren't you rich?' It was not as though the car was chronically unreliable, for once the tendency for the rear-drive unit to tear away from its mountings on the glass-fibre hull had been cured by the provision of heat shields to ward off the heat generated by the inboard discs which was softening the material, the normal production cars provided their owners with very dependable transport.

The answer lies almost certainly in the fact that there was far too much noise and vibration fed into the interior of a car costing nearly £2,000, and that the price itself was somewhat excessive for a 1,300cc two-seater. Originally, it had been intended to develop a special version of the Coventry Climax engine with a cast-iron block, which would probably have reduced both the noise and vibration problems, but this never proved possible and the 'quality feel' of the car suffered thereby. So chronic was the vibration problem that Lucas reckoned an Elite was better than any test rig for vibration testing of such items as dynamos and starter motors.

With the Elite posing its own problems, and with the Formula 1 cars earning a most unenviable reputation for unreliability in Grand Prix racing, Lotus fortunes were at a somewhat low ebb in 1959 as regards both morale and finance—the company made a loss of £29,062 that year. But salvation was already in sight in the shape of the ugliest Lotus racing car ever built, the Lotus 18, which made a most undistinguished first appearance in lowly Formula Junior form at the 1959 Boxing Day Brands Hatch meeting.

The Formula 1 version of the Lotus 18, using the same basic structure as the Formula Junior car, made its first appearance in February 1960 in the Argentine GP. As the car had been completed only after all-night

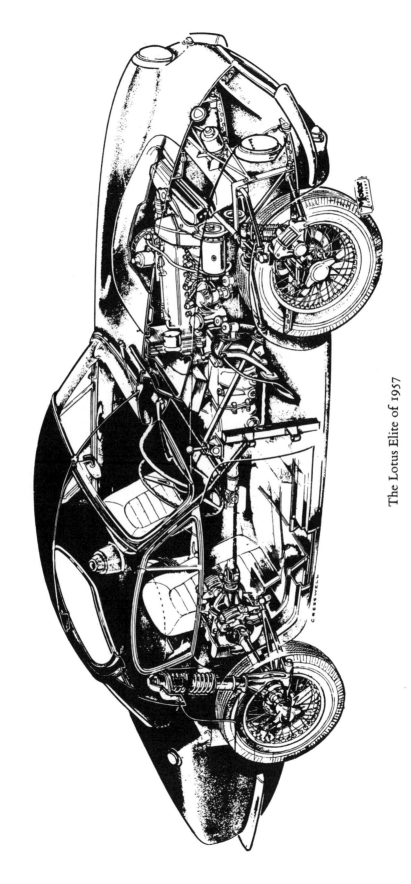

The Lotus Elite of 1957

work at the factory, in which Chapman himself lent a hand, and had then been 'tested' by a hundred-yard dash down the road outside the factory just to ensure that it really worked, not too much was expected of it on this initial outing. But at once it showed far more promise than any previous single-seater, for Innes Ireland, who was then No 1 Lotus works driver, made the front row of the grid with second fastest practice time—Stirling Moss being the quickest—and led the race at the start and again subsequently, but finally finished sixth after a front hub had disintegrated fairly completely.

Ireland subsequently beat Moss at Goodwood and at Silverstone, so Rob Walker bought a Lotus 18 for Moss to drive and with it, at Monaco in May 1960, Moss gave Lotus their first Championship Grand Prix victory, repeating the process in the final Grand Prix of the season, the United States GP at Riverside.

Why had success suddenly come to Lotus? The Lotus 18 was a very simple car, in great contrast to the ill-fated 16, which was diabolically complicated. Chapman followed Cooper's lead and sat the big four-cylinder Coventry Climax engine at the rear of the chassis next to its transmission, thus eliminating those many jointed shafts conveying the drive from the front engine to the transmission and wasting power en route. The square-built, space-frame chassis might have been inherited from a Tiger Moth aeroplane and was about as simple to maintain. The front suspension now employed a normal top wishbone instead of using the anti-roll bar as the front arm, but it was at the rear that the most drastic changes were to be seen, for the Chapman strut suspension had been abandoned in favour of a type of suspension layout developed by Eric Broadley for his Lolas, with a lower wishbone link pivoted at its apex, a fixed-length drive shaft forming the upper wishbone and parallel radius arms pivoting at their forward ends on the bulkhead behind the driver's seat. Where Chapman's layout differed decisively from the Broadley-type rear suspension, however, was in the use of exceptionally long rear wheel uprights of cast magnesium which extended to within four inches of the ground.

This feature of the design was severely criticised when the car first appeared, many critics asserting that should a tyre puncture at speed the bottom of the wheel casting would dig into the road and send the car out of control. In actual fact, when the Rob Walker Lotus 18, with Moss in the cockpit, shed the right rear wheel at 140mph during practice for the Belgian GP at Spa, Moss always reckoned it was the upright that kept the car on an even keel and prevented it from rolling.

The purpose behind this unusual Chapman variation of the Broadley layout was to lower the roll centre substantially at the rear in order to reduce understeer and produce a much more responsive car. On previous Lotus single-seaters, the roll centre had been at hub level or even higher, whereas with the new layout it would be at ground level at front and rear. This had the disadvantage that the roll angle of the wheels was now equal to the roll angle of the car, with a consequent reduction in cornering power, so after some testing on the road, the roll centre was taken up to between 1 and 1½ in above the ground. With its low frontal area, excellent traction and outstanding roadholding, Lotus at last had a Grand Prix car capable of vanquishing the opposition. A two-seater version, the Lotus 19, was equally successful in sports-car racing.

The following year, 1961, brought with it a new Grand Prix formula which reduced engine capacity from 2½ litres to 1½ litres. With less power now available, aerodynamic form once again became most important, and the Lotus 21 Formula 1 car was the first Lotus for many years to house the coil spring and telescopic damper front-suspension units inside the body out of the wind. The upper wishbone was now a fabricated sheet-steel rocking lever pivoted in the middle with its outer end attached to the top of the front wheel upright and its inner end acting on the suspension unit.

At the rear, too, there were changes, for an upper wishbone link had been added to relieve the half-shaft of the strain of acting as a sideways locating link. The half-shafts were now fitted with an inner universal joint with a flexible rubber connection in place of splines. The whole car was much smoother and also lower, for this was the first Lotus single-seater in which the back of the driver's seat was raked to the rear, thereby lowering his position in the car. And the Lotus gearbox, which had never ceased to be a problem, had been replaced by a ZF unit. Colin Chapman always held that designing a gearbox was not difficult, but designing a gear-selection system that really worked was indeed very difficult.

The Lotus 21 had little success, for it was powered only by the 150hp four-cylinder Coventry Climax engine designed some time previously for the 1,500cc Formula 2, whereas the Ferrari were propelled by an excellent V-6 engine developed in Formula 2 races the previous season, and developing around 180bhp. Even so, Moss scored two fantastic and quite unexpected victories with his modified Lotus 18 of the previous year, winning both the Monaco GP and the German GP on the Nürburgring by sheer genius. By the end of the season, however, the new 1½ litre Coventry Climax V-8 was beginning to show promise, and for this engine Chapman designed not one but two cars for the 1962 season. The first, the Lotus 24, was a lowered 21 with the back of the driver's seat even more raked and the space frame around the engine compart-

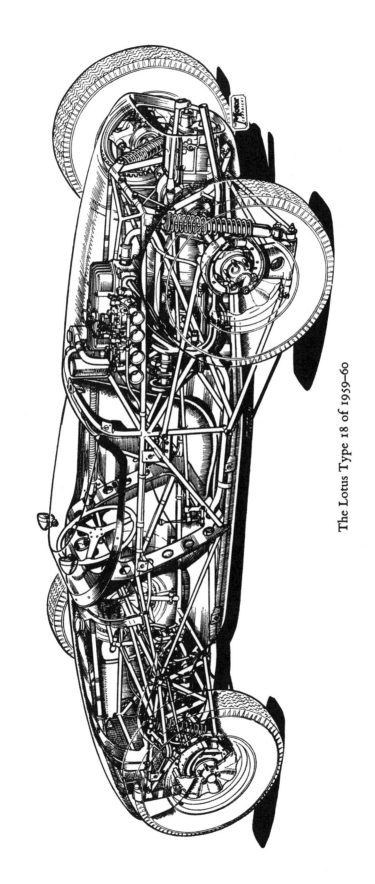

The Lotus Type 18 of 1959–60

ment modified to provide room for the wider V-8. Front and rear suspension were the same as for the 21.

The Lotus 25, on the other hand, was a very different racing car, and one destined to influence all its rivals. It was a monocoque, and was, in fact, conceived in late 1961 before the 24, but the space frame 24 was built around the same running gear as the 25 in case the monocoque car did not work. The main reason for turning to a monocoque was that the Coventry Climax V-8 produced more power and therefore burned more fuel than the four-cylinder engine, and accommodating the extra tankage it required was no mean problem. In fact, the first concept of the 25 was of two long side tanks with the front suspension mounted on the forward ends and the rear suspension on the after ends of the tanks. Moreover, Chapman had been influenced into thinking monocoque thoughts by development work carried out for the next production Lotus, the 26, later to be known as the Elan. In its earliest form, this was to be another self-supporting glass-fibre hull like the Elite, and to test out the running-gear, a backbone chassis was knocked up to run on the road. This backbone chassis displayed such remarkable torsional stiffness that Chapman began thinking in terms of mono-cocques for racing cars. The 25 was not, in fact, a full monocoque, as racing cars so constructed are very difficult to maintain during the season owing to the difficulty of obtaining access to their vitals.

The Lotus 25 made its racing debut in the Dutch GP at Zandvoort in May 1962 where it led the race until clutch trouble made a long call at the pits necessary. Thereafter this car, with Jim Clark at the wheel, dominated the Grand Prix scene during the years of the 1½ litre formula. With it, Clark was runner-up in the World Championship of 1962, losing the championship in the last race of the series when a small bolt came out of the engine in South Africa, won the championship in 1963, lost it again in the final laps of the final race in 1964 to be placed third, and won it for the second time in 1965. From 1964 onward, the 25 was developed into the 33, with a host of minor modifications, but the general concept of the car was the same.

This was the golden era of the Chapman-Clark collaboration, a designer of genius working hand in hand with the finest racing driver of the age, each with the utmost faith and trust in the other. Clark made no pretensions at all to being a designer. During practising, he would come into the pits, tell Chapman what the car was doing wrong, then wait trustingly for Chapman to put it right. And when he was racing, Chapman never tried to control him from the pits, but just fed him with all available information and left Clark to make his own deduc-tions. For Chapman knew that Clark would obtain the utmost from any car that he drove, and in fact time and again Clark won with a car so sick that most other drivers would have given up all hope and merely plugged on to finish.

The year 1962 was indeed a momentous one for Colin Chapman and Lotus, for in addition to the introduction of what is generally agreed to be one of the great Grand Prix cars of all time, the year also saw Chapman realise another project dear to his heart and which he had been nourishing for some time, namely the entry of Lotus into the engine business. Racing Engines Ltd had been formed as one of the many Lotus sub-companies while the concern was still at Hornsey, but now the name came to active life with the announcement of a 1,500cc Lotus engine. This new engine was based on the Ford 105E/109E range of engines designed originally for the Anglia and Classic and later developed by Cosworth Engineering Ltd into the most formidable of all Formula Junior engines. Cosworth was formed by Mike Costin and Keith Duckworth in 1958. Mike Costin remained with Lotus until 1962, before joining the Cosworth concern full time.

The new Lotus engine consisted of the Ford cylinder block to which was added a special light-alloy twin-overhead camshaft head designed by Harry Mundy. The head was notable for its narrow valve angles, both the inlet and the exhaust valves being at only 27 degrees from the vertical in order to get a reasonable compression ratio without danger of the valves coming into contact with the pistons. First appearance of the new Lotus engine was in a Lotus 23 at the Nürburgring 1,000 kilometres race in May 1962. The Lotus 23 was a space-frame two-seater sports-racing-car version of the then current Lotus 22 Formula Junior single-seater, and was notable for its light weight and very smooth shape. It was also notable for its 13in wheels which posed a special problem for the tyre companies, as the power units fitted to the 23 grew steadily in capacity and power from the original 1,000cc Cosworth-Ford engine of 105bhp to 1½ litre V-8 Formula 1 engines.

The debut of the new Lotus engine was indeed sensational. At the end of the first 14 mile lap of the Nürburgring, the little Lotus 23 with Jim Clark at the wheel came through first on its own, so much on its own that spectators assumed some giant pile-up on the first lap had delayed the big Ferraris and the works eight-cylinder Porsches. But not a bit of it, for half a minute later they all came streaming past the pits, and by the end of the first hour Clark had extended his lead to nearly a minute and a half. Then an exhaust pipe came adrift and fumes filled the cockpit, so that when, towards the end of the second hour, the front brakes played up on a bend, a somewhat gassed Clark ran off the road.

A pair of Lotus 23s one month later were at the very centre of the

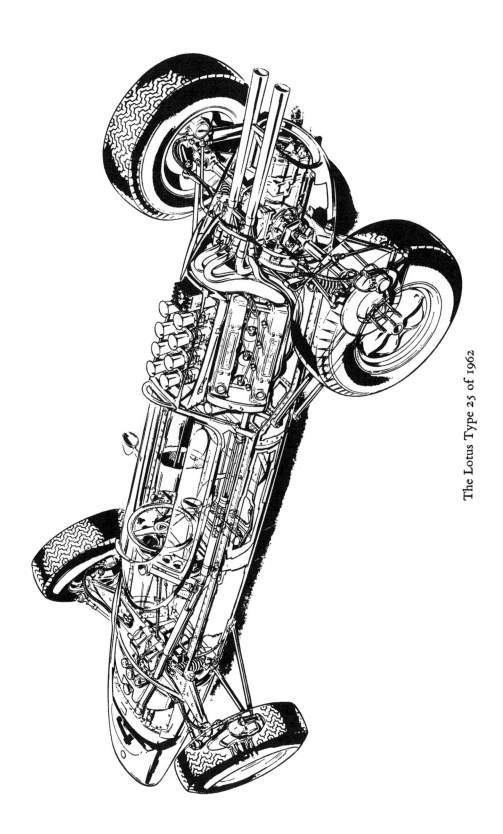

The Lotus Type 25 of 1962

great Le Mans row. When the two cars appeared for scrutineering, they were both turned down on various counts, including insufficient ground clearance, too large a turning circle, petrol tanks of too great a capacity and, most serious, the front and rear wheel methods of attachment were not the same, the rear wheels being held on by six studs and the front by only four, which meant that the spare wheel would fit either the front hubs or the rear but not both. All these matters were attended to and the cars were again presented to the scrutineers, only to be turned down again on the grounds that they did not comply with the spirit of the regulations, a point that might have occurred to the club when they received the entries for the two cars the previous February accompanied by photographs and drawings. A justifiably incensed Chapman made some pretty bitter remarks about the French organisers, who, understanding more English than Chapman gave them credit for, told each other gloomily that Chapman 'was not a gentleman'. Chapman there and then made a mighty vow that never again, so help him, would there be Lotus works entries at Le Mans. Nor have there been.

The new Lotus twin-cam engine was not designed and built primarily to power the Lotus 23. Its true home was destined to be under the bonnet of the Elan, a front-engined open two-seater sports car introduced in October at the 1962 Motor Show. This car was to play a vital part in the future of both Chapman and Lotus, for it succeeded where the Elite had failed, and became a viable commercial proposition, thereby giving Lotus the backing of a true production car.

The Elan was not only significantly less expensive than the Elite—when introduced its basic price was £1,090, whereas the Elite had cost £1,375—but it was also a much more civilised and practical car for everyday use on the road. The basis of the Elan was the welded steel backbone chassis whose immense stiffness and lightness had started Colin Chapman thinking deep thoughts about monococque racing cars. The suspension was conventional Lotus, with coil spring and wishbone link front suspension and a modified version of the Chapman struts at the rear, but provided both roadholding far in advance of all contemporary British sports cars and a ride more comfortable than that of many saloons. The Lotus engine, with its capacity increased to 1,600cc by the time production began in February 1963, delivered 105bhp with great smoothness and flexibility, sufficient to propel the very light and smooth Elan at a maximum of 111.9mph allied with quite outstanding acceleration. It was significant that the introductory party for the Elan was held in the Regent Street showrooms of the Ford Motor Co for, of late, there had been ever-increasing co-operation between Lotus and Ford. A further stage in this co-operation was the introduction, in January 1963, of a Lotus-modified version of Ford's best-selling saloon, the Cortina. One of the main objects of this co-operation was to provide the Ford Competitions Department with a race- and rally-winning Cortina. To this end, the Cortina was endowed with the Lotus twin-cam engine and the rear suspension was drastically modified by substituting coil springs, radius arms and an A-bracket for the normal semi-elliptic leaf springs.

It is only fair to mention that in spite of the troubles that assailed the earlier versions of the Lotus-Cortina, the car succeeded in achieving what it was originally designed to do, namely win the British and European Saloon Car Championships. In order that the Lotus-Cortina should be eligible to take part in these championships, the rules demanded that at least a thousand must be built and the car duly homologated to this effect, and this need for swift production during its early life limited the extent of the modifications that could be made to the rear suspension. These modifications were, in fact, restricted to those that could be incorporated while still making use of the attachment points provided for the leaf-spring rear suspension. If Colin Chapman had been given a free hand, the modifications would have been much more extensive and the troubles that subsequently arose would probably have been eliminated at the design stage.

Chapman's co-operation until now had been with Ford of Britain, but by the end of 1962, he found himself embarked on a joint project with the parent company in the United States. The project was—to win the Indianapolis 500. That Chapman ever became involved in Indianapolis was chiefly Dan Gurney's doing. The American driver was a member of the works Porsche Formula 1 team in 1962, but he had Indianapolis ambitions, thought that a rear-engined car could win, and when he saw the monococque Lotus 25 at the Dutch GP, he reckoned that here was the basis for a rear-engined Indianapolis winner. So he talked Chapman into going to watch the 500 that year, and Chapman returned to England convinced he would have no difficulty in designing a car to beat the antique front-engined roadsters which then dominated the Indianapolis race. If only he could find a suitable engine. With his close liaison developing with Ford of Britain, it was but natural he should look to Ford of America for a big-capacity Indianapolis engine. When Dan Gurney and Chapman flew into Detroit in July 1962 to call on Ford to discuss their project, they found they were talking to the converted, for Don Frey of Ford had also been at the 1962 500, and he had left after the race thinking in terms of a Ford Indianapolis engine based on the Fairlane production engine.

The Ford people at this time were in touch with the established Indianapolis car-builders and knew absolutely nothing about this Eng-

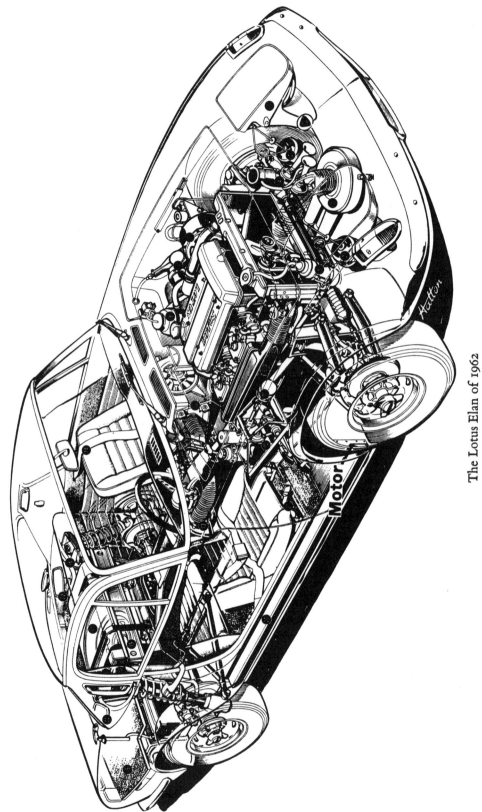

The Lotus Elan of 1962

lish character who stated so confidently that he could build a car to win Indianapolis if Ford would supply him with an engine that developed 350bhp on normal petrol. A mere 350bhp, for most of the four-cylinder Offenhauser engines that were standard power for Indianapolis winners were developing around 400bhp on methanol and nitromethane.

But Chapman reckoned that his single-seater would weigh much less than the front-engined roadsters and would therefore require only one tyre change instead of three or four. Moreover, if it ran on petrol and averaged around 7mpg, it would require only two refuelling stops whereas the much thirstier 4mpg roadsters on their special fuels had to refuel three or four times during the 500 miles.

Chapman's friends speak of his supreme self-confidence; his enemies talk of his arrogance. In any case, it is typical of the man that he did not approach one of the mightiest motoring corporations on earth cap in hand and humbly ask what sort of a car they would have him build. On the contrary, it was Chapman who told Ford what they needed to win Indianapolis, and he stated his terms for building them such a car, terms which in fact meant that Ford paid for the whole operation. The sheer audacity of the man takes one's breath away.

Moreover, having gained Ford's consent, Chapman proved his point by so nearly winning the 500 at his first attempt, in spite of the almost total ignorance of his team of the very special demands made by this unique race on the handling of a car, of terrible pit work and sheer lack of knowledge of the regulations governing the race. As it was, Jim Clark finished a fighting second in the 1963 500 in a finish so controversial that it was debated for months afterward whether or not the rules had been bent to favour the all-American winner. The following year both the Lotus entries were eliminated directly or indirectly by tyre trouble. This would have finished a lesser man than Chapman, but after a stormy scene he still managed to persuade Ford to back him for a third onslaught on the race.

The first Indianapolis Lotus, the 29, was a direct derivative of the Lotus 25 Grand Prix car, but with its wheelbase increased from 7ft 7in to 8ft 0in, its track increased from 4ft 5½in to 4ft 8in and its fuel tankage upped from 26 to 42 gallons. The 1964 car, the Lotus 34, was a development of the previous year's 29, but powered by the twin overhead camshaft Ford V-8 instead of the pushrod V-8. The new Lotus 38 for the 1965 Indianapolis 500 was a considerable departure, for it was of full monocoque construction and therefore much stiffer. The decision to go to a full monocoque was taken chiefly because of the need for a 40 per cent increase in fuel-tank capacity to enable the engine to be run on methanol instead of petrol, the flexible bag tanks being housed in com-

partments running the full length of the hull on each side. And at last
Lotus and Ford hit the Indianapolis jackpot, for Clark not only won by
more than two minutes, but also led for no fewer than 190 of the 200
laps.

Chapman had his resounding failures as well as his equally resounding
successes at this time, however, for in 1964 his design for a big-capacity
Group 7 sports car powered by a 4.7 litre Ford V-8 engine was an almost
total disaster. The Lotus 30 was a most interesting backbone-chassis
design with glass-fibre body, which in many ways resembled an Elan
turned back to front with the engine mounted in the rear prongs of the
backbone frame. But alas, it was designed to run on 13in wheels which
were just too small for the power they had to put down on the road.
Bigger wheels could not be fitted without a drastic modification of the
whole design, for there was not room in the body wheel-arches for 15in
wheels. Moreover, the suspension, which was designed for the 13in
wheels, proved unequal to the stresses fed into it and was all too apt to
wilt under the strain. The failure of the Lotus 30 was a thorough-going
disaster, for this racing version was to be the forerunner of a range of
big-engined production cars.

A development of the Lotus 30 was raced in 1965, but the Lotus 40
was no more successful than its predecessor. The year 1965, in which
Jim Clark won the World Championship of Drivers for the second
time and Lotus the Formula 1 Constructors' Championship for the
second time and Indianapolis for the first time, nonetheless ended in
considerable uncertainty, for it was the last year of the old 1½ litre
formula. In January 1966 the new 3 litre Grand Prix formula came into
operation, but Coventry Climax, who had built the engines with which
Lotus had gained all their Grand Prix successes to date, had announced
that they were withdrawing from motor racing at the end of the 1965
season and would not therefore be building any 3 litre engines. This
decision, taken in February 1965, meant that the Lotus team would be
engineless next year unless a substitute could be found. It is now history
that a substitute was found, thanks to Ford of Britain backing the
Cosworth concern to the tune of £100,000, but the decision to do so
was not taken until October 1965, which meant that the new engine
would not be raceworthy until the 1967 season.

In the meantime, i.e., in 1966, Team Lotus went racing with a 2 litre
version of the Coventry Climax V-8 in the previous year's Lotus 33,
and more spasmodically with a new Lotus 43 powered by the BRM H-
16 engine. This engine was in diabolical trouble throughout the season,
both in the works BRMs and in the Lotus 43s on their infrequent
appearances. Just once the engine worked properly, and ironically it was

Page 381 (above) With Jim
Clark (left) Chapman formed a
driver-constructor partnership
unique in motor racing, a
partnership that was to bring
Clark and Lotus two World
Championships; Innes Ireland
on the right; (*below*)
reluctantly, Chapman abandoned driving himself and
concentrated on team
management. Hazel and Colin
watch anxiously from the pits
at Rheims as Jim Clark reels off
the last few laps to win the 1963
French Grand Prix for Lotus

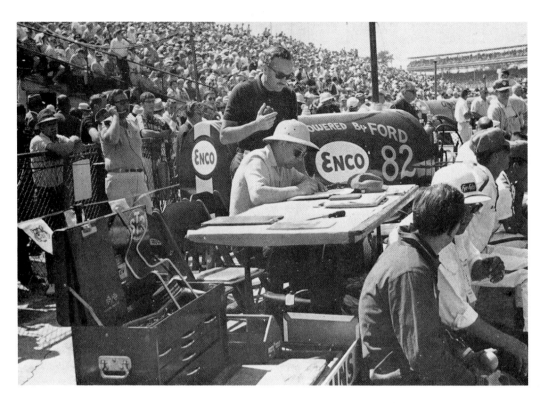

Page 382 (*above*) From the pits, Chapman directs the successful Lotus assault on the richest race in the world, the Indianapolis 500, (*below*) Jim Clark crosses the line to win the 1965 Indianapolis 500, the first British driver ever to do so. That year Lotus were first, second, fifth and eighth

a spare unit borrowed by Lotus from BRM on the eve of the US Grand Prix. It lasted the entire race and gave Clark his one victory of the season. The Lotus 43 in which it ran was a monococque that owed much to the Lotus 28 Indianapolis car, but the monococque finished abruptly behind the driver's bulkhead, the rear suspension then being mounted on the rear of the H-16 engine.

The same form of construction was used for the 1967 Formula 1 car, the Lotus 49, designed specifically to take the new Ford V-8 engine and planned in conjunction with Cosworth, the builders of the engine. Chapman thought that an entirely new engine would provide sufficient problems without adding to them by building a radically different chassis as well, so all thought of four-wheel drive or De Dion suspension was pushed firmly into the background and the new 49 was made as simple as possible and was based on the 43 with which they already had some experience. It is now history that the Lotus 49 and its new Ford engine won on its first appearance the Dutch GP at Zandvoort in 1967. Clark went on to win the British, US and Mexican Grands Prix and to finish third in the Championship of Drivers.

Back at the factory, the works at Cheshunt were becoming ever more congested as Elan production soared from a somewhat faltering five cars a week at the beginning of 1963 to around forty a week by September 1966, in spite of the acquisition of an adjacent factory solely for the production of Elan bodies. Yet despite the outwardly prosperous state of affairs at Lotus, Chapman now realised that the company was not as financially successful at it should be. Production had soared, but most of the money made by the additional sales was being spent in keeping the factory running. Where Chapman differed from so many other automobile engineers of genius is that he realised in time that such a state of affairs existed, and took steps to cure the situation before it became disastrous. He made the usual Chapman intensive study of modern management techniques, applied them to Lotus and the company began to run a substantial profit.

Chapman had for some time been looking for a new site on which to expand—one well away from a centre of population where he would no longer be plagued by constant complaints from the local inhabitants of excessive noise. One of the great attractions of the new site at Hethel, near Norwich, was not only that it provided plenty of space for future expansion, but also that it was an ex-RAF airfield with a perimeter track that could be used by Lotus as their own private test circuit. Moreover, the airfield provided plenty of space for Chapman to land and take off in his own private aircraft which he had been using to an ever-increasing extent for his business journeys. When Chapman first began to fly himself from circuit to circuit in the early 1960s, his flying technique was almost as 'press-on' as his normal road driving—in one journey to Brussels for a Formula 1 race there, he was said to have frightened both a Sabena 707 and diners in the Atomium restaurant when he arrived over Brussels in conditions of somewhat low visibility. He took his flying extremely seriously, however, qualified for all the required licenses, and was apt to spend more time talking about aircraft than about cars.

The foundation stone of the new Hethel factory was well and truly laid by Chapman in July 1966, and in November of that year the whole Lotus factory moved from Cheshunt to Hethel in the course of a single weekend, the shadow Minister of Transport performing the official opening ceremony in August 1967.

To ensure that the new factory was kept fully occupied, two additional Lotus models were introduced, the first, in December 1966, being the mid-engined Lotus 46 Europa and its racing derivative, the 47. The Europa was designed around the Renault 16 engine and transmission and was primarily for the European market, to sell there at a very competitive price. Then in August 1967 came the bigger, four-seater version of the Elan, the Lotus 50 Elan Plus 2, a much more sophisticated machine than the Elan and designed to appeal more to the rising young executive than to the sports-car driver.

The year 1968 was indeed a climactic one for Colin Chapman. In April, Jim Clark was killed in a Formula 2 Lotus at Hockenheim, and I do believe that Chapman would most thankfully have pulled out of racing there and then if he had not been committed to continuing by his contracts. Clark was not only his chief driver, he was also a close personal friend and a collaborator. Colin designed cars for Jim and Jim drove them as well as he could to bring success for Colin. Now, the whole impetus for the time being seemed to vanish from Chapman's life, and motor racing had lost its point.

But Chapman had to go on, for plans laid many months previously were now reaching fruition. These included another Chapman breakthrough—the fantastic wedge-shaped Lotus 56 Indianapolis car with a gas-turbine power unit and four-wheel drive. This, indeed, was the shape of things to come. But although Joe Leonard and Graham Hill placed the new Lotus 56 cars in first and second places on the starting-grid, Leonard's four-lap qualifying speed of 171.559mph and his second lap of 171.953mph beating all previous records, the qualifying trials saw another tragedy for Lotus when Mike Spence, with the third car, brushed the wall and was killed when a front wheel came back into the cockpit. In the race itself, Leonard led at the beginning and was fifteen seconds

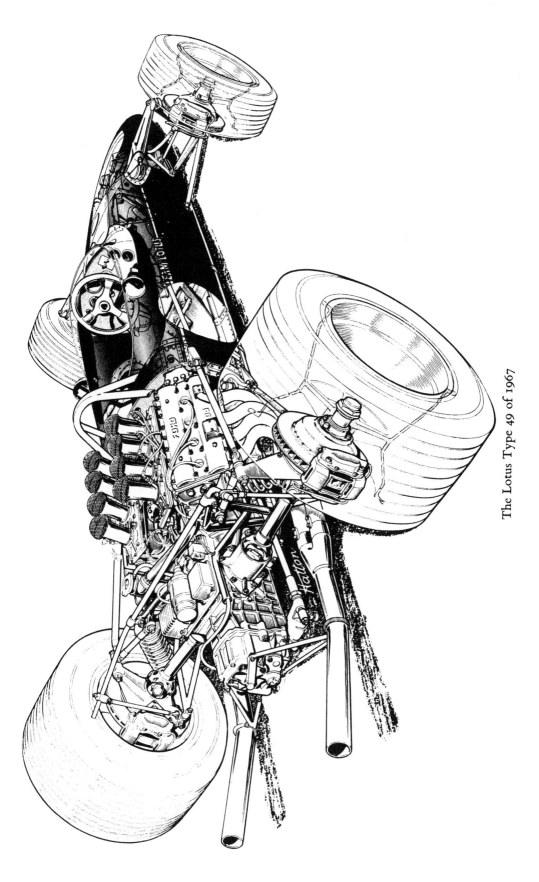

The Lotus Type 49 of 1967

ahead of Bobby Unser, the eventual winner, on the 191st of the 200 laps when the car broke a final drive shaft and was out. A similar fate befell Pollard's Lotus 56, and Graham Hill went out much more dramatically when the right front suspension collapsed and he hit a wall.

In September of 1968 Lotus Cars became a public company, and with their shares offered for the first time on the Stock Exchange at 30 shillings soaring rapidly to 47 shillings and beyond, Colin Chapman became a millionaire at the age of forty. He seemed outwardly much more excited when, from the front of the pits at Mexico, he saw Graham Hill win the Mexican Grand Prix and the World Championship of Drivers, thereby ensuring that Lotus won the Formula 1 Constructors' Cup for the second time. In spite of his Big Business Man status, Chapman became immensely excited by the success of his team and was apt to grapple nearby Lotus mechanics in a bear-like hug and to cast his cap in triumph on to the ground.

On a visit to Hethel, I was discussing the progressive development of the 1½ litre 25 and its development, the 33 with Colin Chapman. 'It's odd,' he remarked, 'but both the 25 and the 33 have down the years become progressively slower in spite of Coventry Climax providing us with more power. Slower in their potential speed, that is. But their lap times have just as steadily improved, so that these slower cars are in fact faster round most circuits. I'm sure the reason is the development in tyres, which have grown steadily wider so that the front ones have significantly increased the frontal area to which these marginally pow-ered cars are very susceptible. But the wider tyres have substantially raised the cornering speed, including improving the traction out of corners.'

This line of thought, that high cornering speeds and the ability to put the power down early and harshly are more important than sheer maximum speed were to influence greatly the design of Lotus Grand Prix cars in the '70s and '80s. The problem of putting the maximum power on to the road had not been all that great with the basically underpowered 1½ litre cars, but the advent of cars built to the new 3 litre formula with even at the start double the horsepower accompanied by only a marginal increase in minimum weight—up from 450 kg to 500 kg—was a very different story. Before long we had Grand Prix cars which were causing the tyre makers considerable anguish by their ability to spin their rear wheels when accelerating hard to third gear and even in top.

That the problem was essentially an aerodynamic problem was shown by tests conducted at Indianapolis in 1967 with a Lotus 38 fitted with an aircraft-type black box recorder which showed an astonishing amount

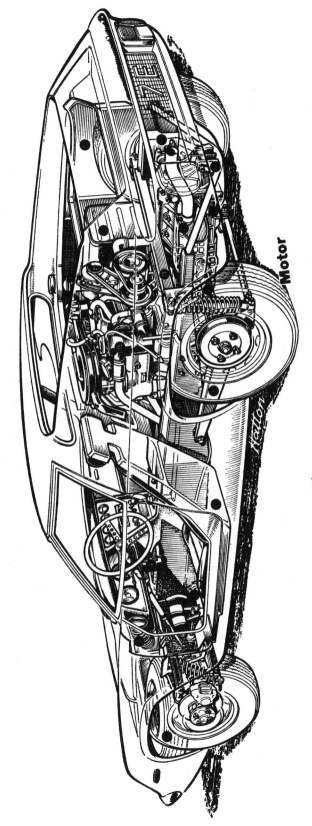

The Lotus Europa of 1966

of aerodynamic lift generated by the car at speed, especially at the rear. Colin's first response to this new knowledge was the wedge-shaped car with bodywork designed, especially at the rear to create its own aerodynamic downforce. The 49B appeared early in 1968 with an upswept tail cowling and small nose fins, but the true wedge cars were the Lotus 56 Indianapolis entries which were gas turbine powered and employed four-wheel drive, and which so nearly won the 1968 Indianapolis 500 from pole position, retiring when in the lead on the 191st of the 200 laps with a broken fuel pump drive shaft.

From this Indianapolis car was developed the Lotus 56B raced in 1971, the only gas turbine powered car to compete in Grand Prix racing. That it was not successful Colin blamed on its four-wheel drive, which also applied to the normally powered—Ford V8—Lotus 63 four-wheel-drive GP car raced in 1969 and which was equally unsuccessful. Colin came to the conclusion that in spite of its apparent attraction for putting the much increased power now available on the road, four-wheel drive in fact worked only for transmitting relatively low torque to the wheels. It therefore suited Indianapolis which was run at a more or less constant speed for the entire lap or for rallying over loose-surfaced tracks or on ice. If, however, more torque was fed into the wheels the driver could no longer control the front end with the steering wheel and the rear end with the throttle because the throttle was affecting both ends, leading to the chronic understeer from which all four-wheel-drive racing cars seem to suffer. Four-wheel drive on racing cars worked only when the torque fed into the front wheels was so reduced that four-wheel drive was no longer worth the added weight and complication.

Chapman therefore decided early on that the only way to increase the cornering power and prevent aerodynamic lift from affecting the traction of the rear wheels was to fit a wing at the rear mounted on the rear hub carriers on which it would then act directly, with smaller side wings at the front to counteract the downforce now being exerted on the rear of the car.

Not until the 1968 French Grand Prix at Rouen, however, did the 49B appear with a massive rear wing mounted on its spindly struts on the rear uprights. Thereafter a contest broke out between the teams as to who could mount the biggest wing. It was soon found, however, that if these big wings were included to exert the maximum downforce for braking and cornering, their drag would slow the car on the straight sections of the circuit to a quite unacceptable extent. So the teams developed various means of tilting the wing for corners then raising it to a near horizontal setting for the straights. Surprisingly, the Lotus wing feathering system was pretty crude, consisting of a pedal linked to the wing by cable which when the driver trod on the pedal on exiting from a corner feathered it against the pull of two strands of rubber which pulled it back to the fully downforce position for the next corner as soon as the driver lifted his foot.

The wing craze became ever crazier in 1969, with biplane and even triplane wings appearing in the early months of the season. Lotus went biplane with a second massive wing mounted on the front uprights just behind the nose wings. But already there was trouble, with wings breaking up in mid race and tearing from their supports to hurtle just over the head of any following driver. Colin, however, had found that each increase in wing area produced a lower lap time, so was tempted to continue increasing the size of the Lotus wings. In the Spanish Grand Prix on the Montjuich Park circuit in Barcelona he widened the rear wings of both Graham Hill and Jochen Rindt's 49s, then added an additional strip at the rear. It was all too much, for in the race first Hill and then Rindt went out of control and crashed badly when one after the other their wings went V-shaped and broke from the reverse thrust imposed when the cars landed after taking off over a hump just past the pits.

Wisely, the CSI banned all wings, though imposing the ban during the middle of practising for the Monaco GP caused an explosion of protests—but mostly from the teams rather than the drivers. This total ban was later modified to permit wings to be used but whose size, positioning and distance from the road was strictly regulated.

The 49 had been designed in the pre-aerodynamic era when frontal area and a nice, smooth shape were the guiding factors—which produced some of the best looking Grand Prix cars of all time. With its add-on aerodynamics it had won the Drivers' Championship in 1968 and 1970 and also the Constructors' Cup in both those years. In 1970, however, it was aided in both these achievements by its successor, which was a superb example of Colin's forward thinking.

The new Lotus 72 of which the design had begun in mid-1969 incorporated the ideas of both Colin and his chief designer, Maurice Phillips. When unveiled to the Press at Brands Hatch in the spring of 1970, it was a strikingly different Grand Prix car. Its wedge-shaped body developed from the Indianapolis cars swept back to a pod on each side of the cockpit into which air was ducted to the radiators for which there was no longer room in the pointed nose and on to an unusual staggered triplane rear wing.

Nor did it just look different. Under the skin were radical suspension and braking innovations. The complicated monocoque centre section was built up from outer panels of soft magnesium alloy 18-gauge sheet formed over steel bulkheads with stiffer inner panels of 20 gauge Alcad. Extending from the front of the monocoque was a subframe of square section steel tubes carrying the suspension, steering and bearers for the inboard mounted disc brakes linked to the front wheels by universally jointed half shafts, the idea being to reduce unsprung weights and improve front end grip. Extending from the subframe was a further light steel tubular space frame carrying the nose inside which were the battery and fire extinguisher. At the rear, the Ford V-8 engine was bolted rigidly to the rear bulkhead of the monocoque and the rear suspension was mounted on a Hewland gearbox.

Compound torsion bars replaced the normal coil springs in the suspension at front and rear. Each torsion bar consisted of a tubular outer section mounted rigidly on the chassis and joined to an inner solid bar whose outer end was linked to its wheel hub by a bell crank so arranged that the system provided rising rate suspension, giving an unusually soft ride which automatically adapted to the change in weight from full to near empty tanks during a race, but with no risk of bottoming with full tanks.

The upper and lower wishbones at front and rear were fabricated from nickel chrome molybdenum sheet. The front wishbones were steeply inclined to provide anti-dive and similarly the rear wishbones were angled to give anti-squat characteristics, features on which Colin was keen for it would mean the angle of the car to the road and therefore the angle of the triple rear wing would not change during hard braking and acceleration.

These three small wings arrayed like a staggered triplane were intended to give the maximum downforce for the minimum drag and to be counter balanced by the top surface of the wedge-shaped nose and its two big nose fins. Moving the radiators into side pods also helped the transfer of as much weight as possible to the rear of the car to aid the use of considerable downforce from the massive wing arrangement.

This was indeed another Chapman original that promised to make all its rivals obsolete. Not surprisingly, however, such a complex new car was not immediately successful. At its April debut in the Spanish Grand Prix on the Jarama circuit north of Madrid, a front brake disc sheared from its shaft, leaving a spinning Rindt with braking on three wheels only. The fault was traced to the discs overheating so ventilated discs were substituted. In the race both cars retired after the soft suspension had allowed them to roll to an unacceptable extent on corners.

The team therefore reverted to the 49s for Monaco, the new cars reappearing for the Dutch Grand Prix at Zandvoort with a stiffer monocoque, thicker anti-roll bars and bigger nose fins. The suspension had also been substantially modified with parallel mountings for the upper and lower front and rear wishbones to eliminate both the anti-squat and the anti-dive characteristics. The anti-squat characteristic had been removed to prevent the tendency of the rear wheels to lift when cornering. The front anti-dive was eliminated at the drivers' request because by stiffening the suspension under braking it removed feel from the already light steering.

Those critics who had been saying ever more loudly that, yes, the 72 was very, very clever but in fact it was just another Chapman failure were silenced when Rindt put the rebuilt 72 on pole position in Holland then proceeded to dominate the race, leading most of the way and winning by 30 seconds after lapping the entire field save for Stewart in second place. After winning four Grands Prix in succession, Rindt and the 72 came to the Italian GP at Monza with a huge lead in both the Drivers' Championship and the Constructors' Cup. Alas, Rindt was fatally injured when during practising the car went out of control under braking when running without its rear wing in pursuit of pole position, but such was his lead in the World Championship that his score was not overtaken and he gained the title posthumously.

For Colin Chapman and Lotus 1970 was, in spite of the tragic death of their leading driver, a most successful year. But 1971 was indeed different. To begin with, Team Lotus did not win a single Grand Prix. As usual, when a team that has been winning suddenly ceases to be successful, there were a number of reasons why the 72 was no longer first past the checkered flag. It had been designed around Firestone tyres that had been made specially to suit its suspension, and it took time for the car's handling to be tuned to the new low-profile tyre Firestone had introduced. Then much of the team's time was devoted to trying unsuccessfully to make the gas turbine powered 56B with its four-wheel drive into a winner. On the driver's side, too, the death of Rindt had left the team with a promising young Fittipaldi whose climb to championship status was delayed by the injuries he received in a road accident and with only young hopefuls to back him up.

The cars themselves were chronically unreliable, retiring time and again, which is often as much a sign of low morale in the racing department when the mechanics feel they are working long hours with no hope of success, as of faults in design.

Chapman himself had matured considerably, for he was now 44 and so was no longer one of the lads. He was a millionaire living a life of considerable style in a superb house and a Jaguar or a massive American Ford were much more his type of car than a Lotus 7. It was not therefore surprising that he decided to take Group Lotus and its production cars up market. The replacement of purchase tax by Value Added Tax meant there was no longer a considerable financial advantage in taking delivery of your new Lotus as a kit of parts, rather than as a complete car at the end of the production line and in any case the role of kit car manufacturer on which Lotus had been built in its early years detracted from the up market image Chapman wished to create.

The major role of Group Lotus and its production cars had from the beginning also been to make money to pay for Colin to go motor racing, which was far more interesting to him than building cars for sale. It is well known that there is more profit to be made in the motor industry from building relatively few high-priced cars rather than a multitude of small, inexpensive cars, which was yet another reason for the shift in emphasis at Lotus.

In pursuit of this up market image only the top of the range Elan Plus 2 which had never been sold in kit form, and the Europa remained in production. Lotus Components, which had become Lotus Racing and built single-seater Formula Ford, Formula 3 and Formula 2 racing cars for sale to private owners was closed, and manufacturing rights to the Lotus Seven were sold to Caterham Cars.

The new four-seater Elite hatchback took the company ever more up market than intended, for Chapman had originally planned it as selling for not much more than the Elan Plus 2 at around £3,500, but further costing on the eve of its introduction in 1974 showed it would have to sell at £5,000. As a result, although production had been planned at the rate of 50 cars a week, it never exceeded 15 a week.

There was even a period of some months in 1974 when both the Elan Plus 2 and the Europa having been phased out the new Elite not yet ready when Lotus built no production cars. What kept Group Lotus afloat financially was a contract to supply Jensen Healey with the new Lotus 907 four-cylinder engine designed for the new Elite. Chiefly responsible for this engine and its development was the ex-BRM Chief Engineer, Tony Rudd, who joined Lotus in 1969.

To the surprise of most people, Colin in 1972 bought two boat building companies, both specialising in highly expensive top of the range motor cruisers. To their production he applied his highly original new vacuum-assisted resin injection moulding process using glass-reinforced plastic for the hull instead of wood. This achieved a weight saving of some 40 per cent. This patented injection moulding system was used later for the production of bodies for the Elite, its derivative the Eclat and the mid-engined Esprit.

On the Grand Prix front, after the appalling 1971 season, the 72 driven by Fittipaldi won the Drivers' Championship in 1972 and the

Constructors' Cup in 1972 and 1973 then continued to soldier on as the Team car until 1975, winning 20 Grands Prix during its long life. Its extended service was due towards the end to the total failure of its designed successor, the 76, which was so disliked by the drivers in the 1975 season that they hurriedly reverted to the good old 72. An interesting innovation on the 76 was an electrically operated clutch controlled by a switch on the gear lever. The driver therefore no longer had to take his right foot off the throttle in order to brake on approaching a corner for he could now brake with his left foot.

In other respects, too, 1975 was a bad year for Lotus for Players reduced their sponsorship to under half of what it had been and Group Lotus made a loss of £488,000.

Nor was 1976 much better, though Mario Andretti who had rejoined the team won the final Grand Prix of the season, the Japanese GP, with the difficult multi adjustment 77 whose track and wheelbase could be changed to suit each individual circuit.

Better times were ahead, however, for while on holiday at his Ibiza home in August 1975, Colin was seized with the idea of designing a car that developed so much negative pressure by the airflow beneath it that it would be sucked down on to the road instead of being forced down by its drag inducing wings.

Tony Rudd was put in charge of the development programme with, as his assistant aerodynamicist, Peter Wright, who with Tony had been exploring a similar idea when they were both together at BRM. Extensive research in the Imperial College's wind tunnel—the only one with a rolling road essential for this line of experiment—was followed by the building of the 78 in which the airflow under the car was funnelled into a venturi to speed it up then allowed to expand as it escaped under the rear of the car. Preliminary tests showed great promise though not sufficient downforce was developed to eliminate wings entirely.

Extra large aerofoil-shaped side pods assisted in producing downward thrust, and at an early state it was found the negative pressure increased markedly the closer the car was lowered to the road surface, and that side skirts were essential to prevent the air from escaping sideways from under the car.

In its first season the 78 won five Grands Prix and was so far ahead of all its rivals that they had to copy it. The following year, 1978, the 78 and its derivative, the 79, were unbeatable, Andretti and Peterson winning eight Grands Prix, Andretti taking the Drivers' Championship and Lotus the Constructors' Cup. The tragedy of the year, however, was the death of Peterson after a multiple accident at the start of the Italian GP.

The attempt in 1979 to build a car that relied completely on its increased ground effect so that wings were no longer necessary was a disaster, the Lotus 80 suffering from incurable aerodynamic problems, so that the team had to revert to the now outclassed 79, other teams having by now built better copies of it. The situation became even more complicated when motor racing's governing body banned sliding skirts which could adjust themselves to maintaining contact with the road surface, the only alternative then being to build the cars with rock hard suspension which with no "give" kept the skirts in contact with the road surface. The use of negative thrust under the cars meant that the faster a car went the greater the downforce clamping it to the circuit, so that cornering speeds rose to fantastic heights as the cars were pressed onto the road by a downforce that could reach three times the car's weight. When this was accompanied by rock hard suspension owing to the latest ban, the cars were indeed most unpleasant to drive, not only subjecting their drivers to very high g forces when cornering but all around the circuit shaking them with the harsh vibration and pounding fed into the car by every slight circuit irregularity.

Colin Chapman's last great breakthrough in Grand Prix design was the Lotus 86 and its development, the Lotus 88, which employed twin chassis, a primary chassis consisting of the side pods connected by the undertray and mounted on small suspension units which went solid on their bump stops as soon as the car picked up speed and the aerodynamic downforces came into play, and a carbon fibre tub mounted on flexible springs to give the driver and major components a smooth ride.

Rival manufacturers, however, realised that if it was successful it would make all the other cars obsolete and it was therefore banned on the grounds that as the springs on the primary chassis went solid as soon as the car moved, all the aerodynamic elements were not suspended, which was illegal. Chapman fought the ban throughout the 1981 season to no avail, the team having to rely on the stopgap 87. In fact when its development, the 91, won the 1982 Austrian Grand Prix it was the first Lotus GP victory since 1978.

It was also to be the last during Colin's lifetime, and was in fact a lucky one gained mainly because the turbo-powered cars that now increasingly dominated Grand Prix racing had mostly retired. The long struggle to develop the twin chassis cars and the fight to have their banning lifted had all proved to be a great waste of time, time which could more profitably have been spent on developing more orthodox Lotus Grand Prix cars. Moreover, all the argument and unpleasantness had soured for Colin the whole racing scene which had been his great delight. Not that he had given it all up, for in his last year he was fully

Page 395 (*above*) The silver-haired Chapman was only 48 when this 1976 Dutch GP pit shot was taken, but the extreme pressure under which he had lived for so many of those 48 years was already self-evident. (Photo courtesy of Geoffrey Goddard); (*below*) The prototype of one of Chapman's great cars, the radical 72 with its wedge-shaped body, radiators in side pods, triplane wings and torsion bar suspension. It gained the Driver's Championship twice, three Constructors' Cups and won twenty Grands Prix. (Photo courtesy of Geoffrey Goddard)

Page 396 (*above*) The Lotus 79 was a refinement of the 78 which introduced ground effect into GP racing, making all rivals obsolete. With Mario Andretti here seen at the wheel the 79 won both the 1979 Driver's Championship and the Constructors' Cup. (Photo courtesy of Geoffrey Goddard); (*below*) The twin chassis Lotus 88B with Mansell driving practised for the 1981 British Grand Prix but was barred from starting in the race by objections from rival teams who claimed it broke GP regulations. The last of the great radical Chapman designs was therefore never able to fulfill its potential. (Photo courtesy of Geoffrey Goddard)

engaged in developing an electronically controlled active suspension system for the next Lotus Grand Prix car. He had also taken steps to make good the Lotus lack of a turbo-charged engine by signing an agreement with Renault for the use of their turbo engine in the 1983 Lotus 93 T for the 1983 season.

Chapman had closed down his boat building companies, but in an effort to make better use of Lotus Research and Development facilities had taken Group Lotus into the consultancy business. A major client was de Lorean and the eventual production car owed far more to Lotus than it did to de Lorean though the tangled financial web in which this company was wrapped is, in 1991, still being disentangled.

Financial problems, in fact, were much on Colin's mind in the 1980s. Group Lotus, with whose production problems he had become so disillusioned that he had moved his office out of the main Group Lotus plant into Ketteringham Hall which housed Lotus R and D and later the racing side, had sold only 345 cars in 1981 and by 1982 were running at a huge trading loss and therefore could not meet the repayment now due of their five-year loan from American Express.

Plans to return to the young executive market (which Lotus now felt they should never have abandoned) with a new less expensive car had led to an agreement with Toyota for the supply of the major components, but Lotus could not raise the money to produce it. It seemed increasingly likely that Group Lotus was heading for receivership, but in fact it was saved when David Wickens of British Car Auctions and Toyota both put money into it, and when later General Motors bought the company.

Sponsorship of the racing team had also become increasingly difficult. John Player had pulled out, their successors Essex Petroleum had collapsed, but then fortunately John Player returned to sponsor the team.

Colin therefore had a great deal on his mind when he flew to Paris on December 15, 1982, for a FISA meeting, returning to the Hethel landing strip late the same night after a very bumpy crossing of the Channel. Early the following morning he had a heart attack and was gone.

What so distinguished Chapman was the breadth of his imagination. Most engineers responsible for racing car design are content to evolve, to improve and to develop. Only Chapman seemed able to conceive a car totally different from all its current rivals. Three times he made all the cars against which his team was competing so obsolete that they all had to be replaced, with copies of the monocoque 25, the aerodynamic based 72 and the ground effect 78. Only the machinations of the authorities controlling racing and his rival team chiefs prevented the twin chassis 88 from making it four in a row.

Looking back over the past decade since Chapman's death, has there been a single instance of the introduction of a new Grand Prix car that was revolutionary rather than evolutionary? I cannot think of one.

Without any doubt Colin Chapman was an arrogant man who believed totally that he was right—well, most of the time. But it was this sheer self-confidence that enabled him to think his own thoughts rather than build look-alike cars. Often aggressive, usually highly suspicious of authority and its petty rules and regulations, Chapman was a man with a very short fuse indeed. Yet within this bristling exterior lurked a man who could charm the birds out of the trees should he so desire, and who was given to unexpected acts of kindness.

He was indeed a hard taskmaster—one famous designer is on record as saying that if you wanted your marriage to break up you went to work for Chapman, for he paid little heed to the time of day—or night—or to the day of the week. But he drove no one harder than himself, more's the pity, for had he not done so he might still be with us to startle the racing world with yet another new and revolutionary concept of the Grand Prix car.

NOTES
ON CONTRIBUTORS

JACQUES ICKX ('The Bollées'): Enjoyed a long life as a freelance motoring writer. His book *Ainsi naquit l'automobile* won him The Pemberton Trophy of the Guild of Motoring Writers.

ANTHONY BIRD ('Frederick Lanchester'): Authored several books including *The Rolls-Royce Motor Car, Roads and Vehicles,* and *Lanchester Motor Cars,* as well as books on antiques, horology and historical biography.

RUDY KOUSBROEK ('Gabriel Voisin'): Dutch author on a wide range of subjects from Western science to 18th century Japan. A personal friend of Gabriel Voisin.

D.B. TUBBS ('Ferdinand Porsche'): Author of many books and translations including *The Age of Motoring* (with Ronald Barker), *The Rolls-Royce Phantoms,* and *Vintage Cars in Colour.* One time Joint Sports Editor and columnist of *The Motor.*

JERROLD SLONIGER ('Hans Ledwinka'): American motoring historian resident in Germany. Contributed to leading Australian and American automobile journals. Author of various books including *German High-Performance Cars.*

MAURICE D. HENDRY ('Henry M. Leland'): New Zealander with particular interest in American cars, and who knew several members of the Leland family. Worked on staff of *Automobile Quarterly.* Contributed to *Car Life, Veteran and Vintage Magazine* and other journals.

MICHAEL SEDGWICK ('Marc Birkigt'): Worked as joint assistant editor of *Veteran and Vintage Magazine,* authored *Cars of the 1930s, Early Cars,* among others, and was from 1958 to 1966 curator of the Montagu Motor Museum.

JOSÉ MANUEL RODRIGUEZ DE LA VIÑA ('Marc Birkigt'): Spanish specialist on Hispano-Suiza.

GRIFFITH BORGESON ('Harry Miller'): American motoring writer resident in France. Contributor to leading motoring journals and author of *The Golden Age of the American Racing Car.*

PETER HULL ('Vittorio Jano'): Formerly an officer of the RAF and currently Librarian of the Vintage Sports-Car Club. Author of *Alfa Romeo* and *The Vintage Alvis.*

ANGELA CHERRETT ('Vittorio Jano'): Since 1965, Angela Cherrett and her husband, Allan, have been Honorary Secretaries of the Alfa-Romeo Section of the Vintage Sports-Car Club, and own a range of Jano-designed Alfas.

RONALD BARKER ('Alec Issigonis'): Motoring journalist since 1955 and formerly a member of the staff of *Autocar* in London. Past-President of the Vintage Sports-Car Club and member of the advisory board of the Montagu Motor Museum.

LJK SETRIGHT ('Dante Giacosa'): Writing professionally since 1961 and independently since 1966, LJK Setright has a background which includes law, engineering, music and much else.

PHILIP TURNER ('Colin Chapman'): Formerly Sports Editor of *The Motor.* Worked as a fitter's mate before going to London University. Served with the RAF 1940-44. Courted his wife in Hornsey, North London, and so saw the beginnings of Lotus.

INDEX

Entries in bold type refer to illustrations